The THINGS THAT NOBODY KNOWS

William Hartston is a Cambridge-educated mathematician and industrial psychologist. In his ill-spent youth, he played chess competitively, becoming an international master and winning the British chess championship in 1973 and 1975. He runs a competition in creative thinking for the Mind Sports Olympiad.

He writes the off-beat Beachcomber column for the *Daily Express*, for which he is also the opera critic, and has written a number of books on chess, numbers, humour, useless academic research and trivia.

?

The

THINGS
THAT
NOBODY
KNOWS

---

*501 Mysteries of Life,
the Universe and Everything*

!

WILLIAM HARTSTON

Atlantic Books
London

First published in Great Britain in 2011 by Atlantic Books, an imprint of
Atlantic Books Ltd.

3 5 7 9 8 6 4 2

A CIP catalogue record for this book is available from the British Library.

ISBN: 978 184887 825 9

Printed in Great Britain by the MPG Books Group
Design: carrstudio.co.uk
Illustrations © Nathan Burton

Atlantic Books
An imprint of Atlantic Books Ltd
Ormond House
26–27 Boswell Street
London
WC1N 3JZ

www.atlantic-books.co.uk

How can we remember our ignorance,
which our growth requires, when we are
using our knowledge all the time?

Henry David Thoreau (1817–62)

CONTENTS

INTRODUCTION

Ignorance, Fruit-Fly Genitalia and
the End of the World

There are known knowns. These are things we know that we know.
There are known unknowns. That is to say, there are things that we
now know we don't know. But there are also unknown unknowns.
These are things we do not know we don't know.

Donald Rumsfeld, 12 February 2002

The trouble with people like Donald Rumsfeld is that they give
ignorance a bad name. The US Secretary of State for Defense was
generally derided when he made the clumsy statement quoted
above, but he was just trying to remember a line from Confucius
quoted by Henry David Thoreau in *Walden* (1854):

> *To know that we know what we know, and that we do not know*
> *what we do not know, that is true knowledge.*

With the wisdom of Confucius supporting him, Thoreau went
on to ask:

> *How can we remember our ignorance, which our growth requires,*
> *when we are using our knowledge all the time?*

While Rumsfeld was simply categorizing different levels of
not knowing, Confucius and Thoreau had a much more positive

9

approach to ignorance, an approach that provides the basic *raison d'être* of this book. I come to praise ignorance, not to bury it; for there is no better key to understanding the vast and ever-growing expanse of human knowledge. The topics covered in the forthcoming pages are exactly what the book says on the cover: things that nobody knows. Many people, when I have mentioned the title of the book, have unjustifiably assumed it to be another of those not-many-people-know-that collections of useless information. It isn't. There may be a great number of such intriguing facts here, but they are only included when they are crucial to explain what nobody at all knows, and why nobody knows it.

More than three hundred years ago, the French philosopher and mathematician Blaise Pascal likened our knowledge to a sphere which, as it grows larger, inevitably increases the area with which it comes into contact with the unknown. Henry Miller put this more succinctly in *The Wisdom of the Heart* (1941):

In expanding the field of knowledge we but increase the horizon of ignorance.

This book is a guided tour around Miller's horizon of ignorance.

When listening to scientists or other experts talking about the latest advances in their fields, I have always found it more intriguing, and generally more enlightening, when they get on to the subject of the things they don't know. Rumsfeld's known unknowns are what determines the direction of future research – and that is what makes ignorance so exciting.

According to a recent estimate in the on-line Ulrichsweb periodicals directory, there are around 300,000 academic journals currently being published around the world. These may come out weekly, monthly or less frequently, but the total number of issues of all these journals in any year must be over 3 million, and with an average in the region of ten papers in each journal, each reporting a previously unknown result, that adds up to over 30 million additions to our knowledge every year, which is more than six every second. There has to be a vast amount of ignorance out there to keep all those journals in material, and the things that nobody knows that I have identified in the pages that follow only scratch the surface.

I ought now to write something about ontology, epistemology, Karl Popper's concept of falsifiability, Thomas Kuhn's paradigm shifts and everything else that contributes to our ideas of reality, knowledge and what is knowable, but there will be plenty of time for that sort of thing later when we get on to the subject of philosophical unknowns. There is, however, just one more subject that I want to mention: fruit-fly penises.

Male fruit flies have tiny hooks and spines on their penises, the function of which – until recently – nobody knew. The standard way to resolve such a question would be to shave these bristles off and see what effect this had on the sex life of the subject. In the case of fruit-fly penises, however, the bristles are so small they can only be seen under a microscope, and even the best scalpel is too clumsy an instrument to attempt to use as a razor. At the end of 2009, however, researchers at the University of California published a paper describing a method of shaving fruit-fly penises with a laser. Not only could they shave off the bristles, but they could even perform the task with such accuracy that only the top third of each bristle was trimmed. By comparing the sexual exploits of unshaven, partially shaven, and totally shaven fruit flies, they could then tell everyone what they wanted to know. Answer: the sole role of the hooks and spines is to act as biological Velcro and keep the male fruit fly attached to the female during sex.

And until the paper was published, that is probably something that even Donald Rumsfeld did not know that he did not know.

After toying with various ways of organizing the material, I finally decided to settle for the most systematically arbitrary of all: alphabetical order by subject. Where appropriate, I have included cross-references to related topics at the end of the subject sections. These are introduced by the words 'see also', followed by the name(s) of the related subject or subjects and the numbers of the relevant unknowns. There are also cross-references embedded within the body of the entries, directing the reader to other entries that shed further light on the topic under scrutiny.

Before diving into the deep end of our pool of ignorance, I cannot resist concluding this introduction with an example of a question we know we can't answer – at the time of writing, anyway. The question is

Will the world end in 2012?

More precisely, the question is whether the world will end on 21 December 2012, a date supposedly predicted by the ancient Mayans. The calculation is based on the Mayan Long Count Calendar, which must be the most complex way of counting our days that humanity has ever devised. Rather than expressing a date in three figures, as the day, month and year (originally chosen to correspond to the period of rotation of the Earth, and the orbits of the Moon around the Earth and the Earth about the Sun), the Mayans used five figures from interwoven counting systems. There were 20 days (called K'in) in a Winal, 18 Winal in a Tun, 20 Tun in a K'atun, 20 K'atun in a B'ak'tun. A Long Count ended after 13 B'ak'tun. Multiply all these together, and you get 1,872,000 days in a Long Count, after which it starts again. That's just over 5,128 solar years, and since the Mayan calendar began on 11 August 3114 BC, the calculations mean that it will reach its end on 21 December 2012 (remember there was no year zero in our calendar).

Actually the Mayans did not predict the End of the World on that date, nor even a great cataclysm, and some say the date had no more significance than any 1st of January, but it's a good excuse for a blockbuster movie, and NASA has been plagued with phone calls from people who believe in it, some even saying that they are contemplating suicide to avoid the horrors that the End of the World may bring.

So the good news is that the world will probably not end in 2012, but we shall definitely know whether the prediction is correct on 22 December of that year.

To be conscious that you are ignorant
is a great step to knowledge.

Benjamin Disraeli, *Sybil* (1845)

AARDVARKS

1. Is the aardvark the closest living relative of a creature from which all mammals evolved?

In 1999 scientists sequenced and analysed the complete mitochondrial DNA of the aardvark, an unprepossessing, somewhat comical ant-eating creature from Africa, whose name is Afrikaans for 'earth pig'. The results showed that the aardvark may be the closest living relative of the ancient ancestor of all the placental mammals – that is, all mammals, including ourselves, apart from marsupials and the egg-laying monotremes (such as the duck-billed platypus). Surprisingly, the genetic make-up of the aardvark is closer to that of the elephant than the South American anteater, which shares its taste in food and its general appearance.

Research suggests that the chromosomes of the aardvark have undergone relatively little change since placental mammals first evolved over 100 million years ago, but how close the first placental mammal was to the aardvark of today is unknown.

Our knowledge can only be finite, while our ignorance must necessarily be infinite.

Sir Karl Popper (1902–94)

AMERICA

2. Who were the first people to populate America and how did they get there?

Until very recently, the so-called Clovis people were thought to have been the original human inhabitants of the Americas and thus the ancestors of all later indigenous people in both North and South America. The Clovis people were named after the town of Clovis in New Mexico where evidence of their existence was first detected by archaeologists in the 1930s. A distinctively shaped spear point found there became the identifying feature of the Clovis culture, and similar items were later found in many other places. The most generally held theory was that the Clovis people had come from Asia some 13,000 years ago, during the last Ice Age, following herds of animals across the land bridge that then connected Siberia to Alaska. The newcomers went on to establish the first human settlements in North America.

The 'Clovis first' theory, has periodically been disputed by claims of finds that may indicate a pre-Clovis population. Most recently, a large hoard of tools and artefacts was unearthed in Texas which appear to date back to 15,500 years ago, some 2,500 years before the Clovis people are thought to have arrived. Furthermore, the existence of huge ice sheets in North America at the time would have made travel by land from Asia unlikely, and the supporters of a pre-Clovis theory suggest that the original inhabitants arrived by sea, probably from Polynesia, arriving in South America and spreading north.

3. Who is America named after: Amerigo Vespucci or Richard Amerike?

For several hundred years, it has been generally assumed that America was named after the navigator Amerigo Vespucci, who in 1499 sailed from Italy on a voyage of discovery to what is now known as Brazil. The earliest known use of the word 'America' is on a 1507 map by Martin Waldseemüller (→ CARTOGRAPHY 70), a map based mainly on information supplied by Vespucci. Yet there is no evidence that Vespucci himself ever claimed to have given the continent its name, and it is known that in later editions of the map, Waldseemüller tried to change the name to *Terra Incognita* ('unknown land').

From the 1960s, however, evidence began to accumulate in support of an alternative theory regarding the origin of the name 'America'. It all began with the discovery of trading records concerning a Welsh merchant, Richard ap Meryk, who had anglicized his name to Richard Amerike on setting up business in Bristol in the late fifteenth century. Salt cod, in those days, was big business, and the Bristol fishermen brought a good deal of it from Iceland until that trade was stopped by the king of Denmark in 1475. They then sought out new fishing grounds, and the records support the idea that they found what they were looking for off the coast of Newfoundland. This discovery, naturally enough, they kept secret, but Amerike is known to have been a major supporter of John Cabot's voyage of discovery to North America in 1497.

It is now known that both Columbus and Vespucci had copies of Cabot's map. The only question is whether Cabot had already named the new land after his sponsor Amerike. Intriguingly, one more piece of the jigsaw has been wedged in to fit that theory: Amerike's coat of arms. This coat of arms includes stars and stripes, and his supporters say that it inspired the American flag. That is possible, but unlikely: the stripes on Amerike's version are vertical, not horizontal; there are only three stars; and as well as red, white and blue, his coat of arms includes a prominent element of mustardy yellow.

4. Did the Chinese discover America before Christopher Columbus?

In 2002 the retired British submarine commander Gavin Menzies published a best-selling book entitled *1421: The Year China Discovered the World*, in which he argues that Chinese explorers not only reached America long before Columbus, but also discovered Australia, New Zealand and Antarctica – and even circumnavigated the globe a century before Magellan. His claim is that fleets of massive junks under the command of the eunuch-admiral Zheng He performed all these feats at the behest of the Chinese emperor. Although some historians denounce his claims as pure fiction, with no evidence to back them, Menzies says they explain some early European maps that appear to give accurate details of lands that were supposedly undiscovered at the time.

In 2006 a map was unveiled in Beijing that had recently been discovered in an antiques shop. The map included Chinese characters stating it was drawn by Mo Yi Tong and copied from a map made in the 16th year of the reign of the Emperor Yongle, which was 1418. This map included Australia and other lands supposedly unknown at the time. Three years later, in 2009, some more Chinese maps came to light. These claimed to be copies of fifteenth-century originals, and had been collected by the late Dr Hendon Harris Jr, who in 1973 had published a book on discoveries supposedly made by early Chinese mariners. Harris went much further than Menzies, suggesting that the Chinese had reached the Americas around 2200 BC and were the ancestors of the Native Americans.

5. What happened to Virginia Dare, the first English child born in the New World, and the Lost Colony of Roanoke Island?

Virginia Dare, born on 18 August 1587, was the first child born in the Americas to English parents, Eleanor and Ananias Dare. She was born into the colony established that year on Roanoake Island, in what is now North Carolina. The settlers, who were sponsored by Sir Walter Raleigh, were led by Virginia's maternal grandfather, John White. Not long after Virginia was born, the colonists ran short of food, and White returned to England seeking fresh supplies and support. But when he returned three years later, the entire colony had disappeared. Before White's departure, the colonists had agreed to carve a cross if they were in distress or under attack, or, if they decided to move the settlement, they were to carve the name of their new location. White found no cross, just the letters 'CROATOAN'. Croatan Island, not far from Roanoke, was the home of the friendly Croatan tribe, but with the onset of equinoctial storms, White was obliged to return to England without ever establishing the fate of his granddaughter, or any of the other settlers.

Theories to account for the disappearance of the colonists range from drowning, cannibalism and Spanish aggression to peaceful assimilation into the local tribes, but their exact fate remains uncertain. The Lost Colony DNA Project is currently trying to compare the DNA of relatives of the Roanoake colonists with DNA taken from people with Native American ancestry to try to determine whether the colonists died out completely, or 'went native' and interbred with the local people.

6. How did Davy Crockett die?

As everybody knows, Davy Crockett died heroically fighting the Mexicans under Santa Anna at the Battle of the Alamo in 1836. Or did he? There are two very distinct versions of Crockett's death:

(i) According to a black slave named Ben who cooked for Santa Anna's forces, Crockett's body was found at the Alamo, surrounded by at least sixteen Mexican corpses, with Crockett's knife deeply embedded in one of them. That seems to tally with the usual story.

(ii) According to other accounts of the battle, around half a dozen Texans surrendered to the Mexicans and were promptly executed by Santa Anna. Some say that Crockett was among them. This version is supported by the memoirs of a Mexican officer named José Enrique de la Peña, who asserted that Crockett did not die in the battle. The authenticity of these memoirs has been disputed.

7. Did Custer's Last Stand ever really take place?

What really happened on 25 June 1876 at the Battle of the Little Bighorn, where General George Custer and his men were wiped out by Chief Crazy Horse and his Sioux braves? The usual tale highlights Custer's heroism when, heavily outnumbered, he and his men shoot their horses and pile them into a barricade (leave out the shot horses if this is being filmed for purposes of family entertainment) and withstand the Red Indian hordes until they are all killed. Since all Custer's 210 men were wiped out, however, all accounts of his Last Stand have come from the other side, and all the early accounts were made at a time of delicate negotiations between the Sioux and the US government, when there were advantages to be seen in portraying Custer in as heroic a light as possible.

Investigations of what is now known as Custer Hill have led to strong disagreement about what happened. The large number of bodies found there, together with other evidence, has been taken by some to support the story of the barricade and the hopeless but heroic Last Stand. Analysis of the positions of spent cartridge cases, however, has suggested to some a picture of men in a panic, running and shooting wildly in all directions, including into the air and into the ground. Later accounts by participants on the winning side have also suggested that it was all over quite quickly, 'in the time it takes a hungry man to eat a meal', as one is quoted as saying.

ANCIENT HISTORY

8. Did Atlantis ever exist?

In the fourth century BC, the Greek philosopher Plato wrote of the lost city of Atlantis, an island that 'disappeared into the depths of the sea in a single day and night'. He placed it somewhere around the Straits of Gibraltar, and the legend of Atlantis has been with us ever since. Historians have generally given little credence to the tale, pointing out that the invention of imaginary cities was a common literary device of Plato's time – but that has never stopped speculation and occasional expeditions in search of Atlantis.

In 2009 there was a report of images of a vast rectangular grid on the Atlantic seabed, which Atlantis-lovers saw as evidence of a lost city. Unfortunately, closer examination suggested that the grid was an image created by the ship that conducted the survey. More credible was a recent survey by archaeologists and geologists of the marshlands of the Doñana Park near Cadiz, Spain, using deep-ground radar, digital mapping and underwater technology. This revealed what could be a city buried in mud by a tsunami. That interpretation is reinforced by the discovery of sites said to be 'memorial cities' built by the survivors. This is hardly the city under the sea described by Plato, but he places its destruction at around 10,000 years before his own time, so the Spanish site, which dates from 6000–5000 BC, could be the origin of the Atlantis legend, even if it was several thousand years later than Plato's estimate.

9. What did the people of the Magdalenian culture, who lived in Western Europe around 15,000 years ago, do with the cups they made from human skulls?

In the fifth century BC the Greek historian Herodotus, in his description of the Scythians who lived on the far side of the Black Sea, relates how they drank from the skulls of their enemies. There have been similar accounts from other cultures, but there was little material evidence until archaeologists investigating Gough's Cave, a Palaeolithic (Old Stone Age) site in Somerset, uncovered fragments of both human and animal bones – including forty-one pieces of human skull. These pieces, when pieced together, were found to be from half a dozen individuals, and showed what the researchers described as 'meticulous shaping of cranial vaults': the skulls had been worked into the shape of cups. However, the archaeologists were unable to tell whether these 'cups' were actually used for drinking, or whether they played a part in some other ceremony, such as a burial ritual.

10. What did the Minoans call themselves?

From around the twenty-seventh to the fifteenth century BC, the Minoans on the island of Crete were one of the world's most advanced civilizations. Their buildings, their art (which influenced that of both Greece and Egypt) and their ability to recover from natural disasters such as earthquakes and volcanic eruptions all attest to a high level of organization and administration. Yet their ethnic origins and language remain unknown, and their writings, in the script known as Linear A, have yet to be deciphered. All of which contribute to the fact that we do not even know how they referred to themselves. It was certainly not 'Minoan', a term invented by the

British archaeologist Sir Arthur Evans after the mythical King Minos of Crete, who kept the Minotaur in his labyrinth – a story possibly inspired by the elaborate cellars of the Palace of Knossos, which Evans excavated in the early years of the twentieth century.

See also WRITING SYSTEMS 494

11. Who was the victor at the Battle of Kadesh?

The Battle of Kadesh, fought around 1274 BC, was one of the greatest battles of history, and was said to have involved more chariots than any other battle, before or since. Kadesh is also the first major battle for which we have detailed accounts from both sides. In fact, it could be said that we know almost everything about the Battle of Kadesh except who won.

The battle was fought between the armies of the Pharaoh Ramses II of Egypt and Muwatallis, king of the Hittites. After misjudging the closeness of the Hittite forces, Ramses allowed his own troops to become split, leaving him vulnerable to a sudden ambush by the Hittites. By his own account, he was on the verge of defeat when reinforcements arrived and drove off the enemy. Both sides then retreated and a truce was signed shortly thereafter.

For the rest of his long reign, Ramses proclaimed Kadesh as a great Egyptian victory, while the Hittites were firmly convinced they had won the battle. Archaeological investigations have failed to produce evidence to support either side's claim.

12. Did the Hanging Gardens of Babylon ever exist?

According to legend, the Hanging Gardens of Babylon, one of the Seven Wonders of the Ancient World, were built around 600 BC by King Nebuchadnezzar II of Babylon for his wife, Amytis of Media, in Iran, who was pining for the trees and plants of her homeland. The Gardens were written about and highly praised by Greek historians of the first century BC, which was about a hundred years after they were said to have been destroyed in an earthquake. The earlier Greek historian Herodotus, who lived in the fifth century BC, is said to have included the Hanging Gardens in his own list of the Seven Wonders, but this list has not survived and there is no definite reference to the Gardens in any of his known writings. Curiously, neither is there any known reference to them in Babylonian writings of the time.

Since the site of Babylon was rediscovered in the nineteenth century, archaeological excavations have produced some evidence that match parts of some descriptions of the Hanging Gardens, but none of this evidence is sufficient to confirm their existence. One suggestion is that the Gardens never existed, but were just intended as a poetic device. Another suggestion is that they did exist, but were in Nineveh, not Babylon, having been built by Sennacherib of Assyria in the seventh century BC. The oldest of the Seven Wonders would then have been a confused amalgamation between Sennacherib's real gardens and Nebuchadnezzar's mythical version.

13. What caused the collapse of many civilizations around the eastern Mediterranean between the late Bronze Age and early Iron Age?

Between around 1200 and 1150 BC, as iron began to replace bronze as the favoured material for tools and weaponry, a number of civilizations around the Eastern Mediterranean suffered cataclysmic declines from which they never recovered.

In Greece, the great stone palaces of the Mycenean culture were all destroyed, in Egypt the period of the New Kingdom came to an end as the country reeled under foreign invaders such as the mysterious 'Sea Peoples', while in the Near East the Hittite empire fell apart, and cities across the region were sacked or burnt to the ground. Various causes, both natural and human, may have lain behind such widespread collapse. There is some evidence of prolonged drought, and of earthquakes and volcanic activity, while some scholars have suggested that mass migrations (possibly connected with climate change) combined with the new iron-based weapons technology may have led to a heightened mood of militarism and a desperate drive for conquest. Or it could be simply that the civilizations that had emerged in the region over the previous two millennia had sown the seeds of their own downfall by becoming too complex to be sustained by the existing systems of rule and administration.

14. What was the original purpose of Stonehenge?

About 5,000 years ago, on Wiltshire's windswept Salisbury Plain, the ancient inhabitants of Britain built a henge, a simple structure consisting of a bank, a ditch and some diggings known as the Aubrey holes. These holes, named after their discoverer, the seventeenth-century antiquarian John Aubrey, are round pits in the chalk, each about 1 metre (3 ft) wide and 1 metre deep, with flat bottoms. Together they form a circle some 87 metres (284 ft) in diameter. Some cremated human bones have been found in the chalk.

The construction was then abandoned for about a thousand years until 2150 BC, when some eighty-two massive bluestones from the Preseli Mountains in southwest Wales were erected on the site. The stones, weighing up to 4 tonnes, would have had to travel some 380 km (240 miles) from their original location, and much research and speculation has been spent on the question of how they were moved. The most likely method involved moving the stones on boats by river and sea, and then using rollers to move them across land. It has also been suggested that the stones may have been transported from Wales by the ice sheets thousands of years earlier, during the last Ice Age.

But what was it all for? It has been suggested that Stonehenge was a temple, or an astronomical observatory, or a centre of healing, or a place of human sacrifice – but with little known about the life or beliefs of the pre-Celtic inhabitants of these islands, there are few clues to go by.

15. What was the origin of the 260-day Tzolk'in calendar of the Mayans?

We have already encountered the ancient Mayan calendar in connection with whether the world will end in 2012 (→ INTRODUCTION), but the reasons behind its complex interweavings of numerical patterns are almost unfathomable. The combination of two different types of week, one of 13 days, the other of 20, to give a 260-day cycle running in conjunction with the 365-day year, is particularly baffling. The commonest – but by no means satisfactory – explanation offered is that the numbers 13 and 20 appear to have held some special significance for the Mayans.

It has been suggested that the 13-day week may relate to the lunar calendar, being the period between a new moon and a full moon. The trouble with this theory is that it doesn't add up, as it would give a lunar month of 26 days instead of the more accurate figure of 29. Supporters of the theory respond by saying that you can enjoy the full moon on the day before and the day after too, so the full cycle is 13 days from new to full, 3 days of full-moon watching, then 13 days of a waning moon. Hey presto: 29. But even the Mayans must have felt there was something unsatisfactory about a 13-day lunar week, with the phases of the Moon starting 3 days later every cycle.

As for the 260-day cycle, one suggestion is that it is the period of human gestation, or at least the time between the first missed period and childbirth. Yet there is no evidence to suggest that Mayan midwives had a large influence on the calendar system.

16. Where did the Etruscans come from?

The Etruscans – the ancient inhabitants of Tuscany – were a major power in central Italy from the beginning of the Iron Age to the early days of the Roman empire. Indeed, it was the Etruscans who were largely responsible for turning the small village of Ruma on the banks of the River Tiber into the mighty city we know as Rome. Yet where the Etruscans came from has been a matter of dispute for more than two millennia. The ancient Romans maintained that their origins lay in Asia Minor; the ancient Greeks, on the other hand, believed they were an indigenous Italian race. The Etruscans themselves left no literature, no religious texts nor any other clues as to their origins, apart from some items found in graves and tombs. What little remains of their language makes it clear that it is not Indo-European – indeed, no similarity has been detected with any known language, alive or dead.

17. When did humans discover that the Earth is round?

'They all laughed when Christopher Columbus said he thought the Earth was round.' Those lyrics from the Ira and George Gershwin song have a good deal to answer for. By the time of Columbus, we had known for around two thousand years that the Earth was round. The Greek mathematician Pythagoras postulated a spherical Earth in about 600 BC and another Greek, the astronomer Eratosthenes, may not have been far out in his calculation of the radius of the Earth around 240 BC – and from that time, no reputable Greek thinker ever suggested that the Earth was anything but round. A few eccentric early Christian theologians reverted to a Flat Earth theory, on the grounds that the Greeks were pagans and therefore must be wrong, but they were always very much in the minority.

The ancient Greeks themselves disagreed as to who was the first to confirm the shape of the Earth, and we lack sufficient knowledge of the astronomical techniques of the ancient Greeks to make any kind of informed speculation. According to Diogenes Laertius, writing in the third century BC, Pythagoras was the first to write of a spherical Earth; according to Theophrastus, it was the philosopher Parmenides in the fifth century BC; and according to Zeno, it was the poet Hesiod around the start of the seventh century BC. However, none of these writers inform us of the grounds on which they make their claims.

18. What is the story behind the thousands of huge jars at the Plain of Jars in Laos?

In north-central Laos, in the province of Xieng Khouang, thousands of huge prehistoric stone jars have been found at about ninety sites, with between one and four hundred jars at each site. Each jar is up to 3 metres (10 ft) in height and about 1 metre (3 ft) in diameter.

When the jars were first investigated in the 1930s, they were thought to be connected to burial practices, as they were similar to other,jars found in Indo-China that had definitely been used for that purpose, but no human or animal remains were ever found in or near the Laotian jars. A local belief is that the jars were used for brewing alcohol, but there is no evidence to support that idea. The jars also seem to be designed to be fitted with lids, and although such lids have been found nearby, no jar has been discovered with the lid in place. Even the age of the jars is unknown, though they are thought to date from the Iron Age, some time between 500 BC and AD 500.

19. What was the cause of death of Alexander the Great?

Alexander III of Macedon was undoubtedly among the most successful military commanders of all time. By the age of thirty, after a ten-year series of campaigns against the Persians and others, he had created one of the largest empires in history. But before his thirty-third birthday, he was dead.

All we know of his death, which occurred in Babylon in June 323 BC, is that it followed an intense fever, and that two days before he died, his soldiers marched past him in tribute as he waved silently. Later Greek and Roman historians, rather like a good many modern journalists, were disinclined to let the facts get in the way of a good story, and came up with a number of different scenarios to account for his death.

Plutarch mentions a fever, which he says developed a fortnight earlier after Alexander had dinner with one of his admirals and then indulged in a drinking session with a friend. Diodorus says he died in agony after drinking a large bowl of wine in honour of Hercules. Others suggest that he was poisoned by Antipater, one of his own generals, who had recently been dismissed as viceroy of Macedonia. Antipater's son Iollas was Alexander's wine-pourer, so would have had both the motive and the opportunity.

A less dramatic explanation is that Alexander died from a combination of heavy drinking and a series of wounds sustained in battle, while one more recent suggestion is that he was poisoned by excessive amounts of the hellebore in the medication he took for his injuries. Death by natural causes from diseases such as typhoid, malaria and West Nile fever are also possible. The latest theory, proposed in 2010, is that Alexander's symptoms – which included not only fever, but also excruciating pains in his liver and his joints, and loss of the power of speech – were consistent with poisoning by calicheamicin, a highly toxic substance produced by certain soil bacteria. Calicheamicin is found in the River Mavronéri in the Peloponnese, a river that the ancient Greeks identified with the River Styx, the mythical entrance to the Underworld, whose waters were said to be deadly poisonous.

20. Were any of the crystal skulls in museums made in ancient times?

The release in 2008 of the film *Indiana Jones and the Crystal Skull* renewed public interest in such skulls, examples of which are on display in some of the world's most reputable museums. Said to have been produced by the long-lost Mesoamerican civilizations of the Aztecs or the Mayans, and believed by some to possess mystical properties, the skulls caught the imagination of the New Age movement in the 1960s – and this no doubt boosted the already flourishing trade in fake relics allegedly from pre-Columbian times, a trade that has gone on since at least the middle of the nineteenth century.

Attempts to date the crystal skull in the British Museum have been made at various times since 1950, and in 1996 a joint study of the BM skull and a similar one in the possession of the Smithsonian Institution in Washington, DC, revealed tool marks that must have been made by a jeweller's wheel – a tool unknown to the Aztecs or Mayans, and which only appeared much later in Europe. The BM consequently reclassified their skull, which they had acquired in 1897, as 'old' rather than 'ancient'.

Whether any of the skulls in other museums or in private hands are genuine antiquities remains an open question.

See also CANNIBALISM 68, CLEOPATRA 84–6, DRUIDS 139–40, EGYPTOLOGY 158–62, THE GREEKS 217–18, LANGUAGE 255–7, THE PYRAMIDS 396–8, ROME 412, THE SPHINX 440–44, UNICORNS 457, WRITING SYSTEMS 493–4, 496

ANTARCTICA

21. Who was the first person to set foot on Antarctica?

The ancient Greeks named the Arctic after *arktos*, the Greek word for 'bear', referring to the Great Bear constellation, Ursa Major, which is seen in the northern sky. With admirable logic, they called the other end of the globe *Antarktike*, because it was opposite (*anti-*) the Arctic. In the eighteenth and early nineteenth centuries, hundreds of expeditions sailed south for the purpose of fishing or exploration, until the ice stopped them making further southward progress. It was only with the United States Exploring Expedition of 1838–42, led by Charles Wilkes, that the existence of land beneath the ice was confirmed and Antarctica was shown to be a true continent.

The first person to set foot on that land after Wilkes had confirmed its existence may have been a member of the crew on an expedition led by the French explorer and sea captain Jules-Sébastien-César Dumont d'Urville; this may have occurred on 20 January 1840. There is some evidence, however, that the American sealer John Davis may have set foot on the Antarctic Peninsula in 1821, but even he was not sure whether he landed on the continent itself or a nearby island, and the precise location of his landing was not properly recorded. There are similar doubts about the location of the 1840 landing by d'Urville's expedition.

22. What creatures live in Lake Vostok in Antarctica?

Around 4 kilometres (2.5 miles) below the surface of the Antarctic ice lies Lake Vostok, the largest of the subglacial lakes of the southern continent. It has lain hidden for at least 14 million years and possibly twice as long. Its existence was not even suspected until 1967, and not confirmed until 1993. Not even a water sample has been extracted from it, but a Russian team has been drilling through the ice and almost reached the lake when the weather forced them to give up in February 2011. When the coldest season is over, however, drilling will resume and we may soon learn the nature of the life-forms that have grown in this vast but isolated lake, which measures 250 by 50 kilometres (150 by 30 miles). The results will be of particular interest to scientists looking for life elsewhere in the Solar System, as the conditions in Lake Vostok are thought to be similar to those found on some of the moons of Jupiter and Saturn.

Quite apart from the possibility of the discovery of new life-forms in Lake Vostok, there is another huge unanswered question about the lake.

See also CARTOGRAPHY 71

23. What is the cause of the huge imbalance in the Earth's magnetic field to the north of Lake Vostok in Antarctica?

Following the confirmation of the existence of Lake Vostok, a good deal of research on its size and nature was conducted by means of radar, either from the air or on the ground. These surveys revealed the unexpected existence of tidal currents and pockets of warm water, and these suggested both geothermal activity and more than one subterranean source for the waters of the lake. The most surprising discovery was made in 2003, when a large discrepancy was found in the Earth's magnetic field over a considerable area of the lake. The difference between the measured value and the expected value is much greater than can be explained by normal daily variations of the field, and the discovery was seized upon by conspiracy theorists to support a wide range of increasingly bizarre ideas.

Some said the disparity was evidence of a secret city beneath the Antarctic ice. Could it be, the conspiracy theorists speculated, the lost city of Atlantis (→ ANCIENT HISTORY 9), or a US or Russian nuclear facility, or a crashed spacecraft – or even 2 million descendants of Nazis who had fled there after the Second World War?

The most likely explanation, however, is that the disparity is evidence of a thinning of the Earth's crust beneath the waters of the lake caused by unexplained geological factors in the planet's distant past. The project of drilling down to the lake's surface has already taken more than fifteen years, and we may have to wait some time before finding out what is going on at the bottom of the lake, whether it is Atlantis, a Nazi colony, alien activity – or just an interesting piece of geology.

See also PENGUINS 359

ANTHROPOLOGY

24. Is there any biological reality to the idea of different human races?

As our knowledge of genetics has grown, the concept of 'race' in human beings has become ever more difficult to define. Before we knew about evolution and genes, it seemed obvious that human beings belonged to various different races. You only had to look at them. As we learned about genes controlling different aspects of a person's appearance, however, the idea of 'race' became ever more difficult to sustain as a biological reality rather than a social construct or a pseudo-scientific attempt to justify xenophobic prejudice. From the genetic point of view, physical differences such as skin colour or hair texture are very superficial.

Recently, there have been attempts to justify a scientific concept of race based on the idea of breeding communities that remain essentially isolated from other such communities and therefore may develop their own genetic strains over a large number of generations. Opponents of that idea, however, suggest that interbreeding has always taken place, and any idea of 'pure' races evolving would have been scuppered by the historical movements of populations.

25. Why do Native Americans have grooves on the backs of their teeth?

For more than a century, American dentists have commented on a curious feature found in those of Native American descent: their front teeth have grooves on their backs. Such a characteristic has also been identified in Siberians, which has been taken to support the theory that the early inhabitants of North America arrived during the last Ice Age from Asia across a land bridge to Alaska (→ AMERICA 2). It is still very much an open question how and when this tooth-ridge evolved and what evolutionary advantage it could possibly have conferred .

26. Why are West Indian men three times as likely as white Englishmen to contract prostate cancer?

Many recent studies have confirmed that the incidence of prostate cancer and the associated mortality rate among Afro-Caribbeans is significantly higher than in other groups. Some recent research has extended its scope to show that African-Americans and men in West African nations historically associated with the transatlantic slave trade also have high rates of prostate cancer, suggesting that there is a genetic predisposition to it among these groups. Other research – in Guadeloupe, Martinique, Jamaica and elsewhere – has put the blame on diet or on pesticide use.

27. Did the Dogon people of Mali possess inexplicable astronomical knowledge?

From the 1930s until the 1950s, the French anthropologist Marcel Griaule studied the Dogon people of Mali and in 1946 reported that they apparently possessed extraordinary astronomical knowledge, mostly relating to the star Sirius. According to Griuale, they knew it was part of a binary star system whose companion took 50 years to complete an orbit. Sirius, however, is extraordinarily faint, and its companion star is a white dwarf which is completely invisible to the human eye, and whose existence has only been confirmed by mathematical calculations of the orbit of Sirius. The Dogons also apparently knew about the rings of Saturn and the moons of Jupiter.

More recently, doubt has been cast on the Dogons' astronomical knowledge, with another researcher suggesting that they are very vague about which star they are referring to. All the same, Griaule's accounts remain perplexing.

If they really did know about Sirius, there are two theories, one considerably more probable than the other. The first suggests that the Dogon learnt about it from alien visitors, presumably from Sirius. The more likely explanation is that they gleaned the information from a team of astronomers who visited Mali in 1893 to see a solar eclipse.

28. What happened to the Khazars?

We have already mentioned the unexplained disappearance of numerous civilizations at the end of the Bronze Age (→ ANCIENT HISTORY 13). A more recent mystery concerns the fate of the Khazars. The Khazars were an agglomeration of various nomadic peoples, who, between the sixth and eleventh centuries AD, coalesced to create one of the largest states in Eurasia, extending across the steppes of southern Russia from the Aral Sea in the east to the Black Sea in the west, and south across the Caucasus to the borders of what are now Turkey and Iran. Towards the end of this period, the might of the Khazar empire was sapped by battles with Svyatoslav of Kiev, then with the Mongol hordes, and their power faded away. Yet over the next two centuries, reports of Khazarian communities and individuals showed that the people had survived, if not their empire.

Many of these reports were from Jewish sources, which is not surprising, as Khazar royalty and most of its aristocracy had converted to Judaism in the eighth century. As a result, various writers (notably Arthur Koestler in his 1976 book *The Thirteenth Tribe*) have speculated that Jewish communities in both Russia and Poland may have descended from the Khazars. This theory has yet to be supported by genetic evidence (→ JUDAISM 253–4).

ARMADILLOS

29. Why do nine-banded armadillos suffer from leprosy?

It is often stated that the nine-banded armadillo is the only animal other than the human that can suffer from leprosy. That is not quite true, as mice and rhesus monkeys have also been infected with leprosy, but the armadillo is certainly the most useful experimental animal for leprosy research, as up to 5 per cent of wild armadillos are thought to suffer from the disease. They are thus not only a valuable source of the bacteria that cause the disease, but also useful subjects for testing possible drugs and vaccines. The question as to why humans and armadillos should have evolved to share a particular susceptibility to the disease may possibly be illuminated when the complete genome of the nine-banded armadillo is unravelled.

AUSTRALIA

30. When did human beings first reach Australia?

The history of *Homo sapiens* – modern humans – is generally believed to have begun in Africa around 200,000 years ago, before our species gradually spread across the rest of the world. We know that the ancestors of the modern Aborigines first reached Australia from Asia, but there is still a large discrepancy between various estimates as to when this happened. The earliest human remains in Australia are from a site at Lake Mungo, New South Wales, and have been dated to around 50,000 years ago. Caution has been expressed about this figure, however, as some say that the carbon-dating techniques used are unreliable beyond 40,000 years.

Rocks bearing Aboriginal art have been dated even earlier, to 60,000 years ago, but with equal caution, and claims of 70,000 years have been made for a discovery of Aboriginal tools. It has even been suggested that an increase in the extent of fires in Australia 120,000 years ago is evidence of human activity at that time.

An associated problem is the question of *how* the first Australians reached their destination. The usual explanation involves a land bridge from Asia to the prehistoric continent of Sahul, formed by what are now Australia and New Guinea. The existence of a land bridge, however, is difficult to reconcile with the lack of similarity between animal species in Australasia and Southeast Asia (the so-called Wallace Line, dividing the fauna of the two regions, cuts through the islands of the Indonesian archipelago). Furthermore, the date at which the land bridge disappeared may not tally with the time the first Australians arrived – so they may have arrived by sea.

31. What killed off the giant kangaroo in Australia?

The giant kangaroo, which was up to 3 metres (10 ft) tall and weighed around 200 kg (450 lb), became extinct around 45,000 years ago, which tallies quite well with theories regarding the date that humans first arrived on the continent (→ 30). A natural conclusion is that the animal was simply hunted to extinction. An alternative theory blames climate change, pointing out that many other large Australian species, including 2-tonne wombats and 5-metre (16 ft) land crocodiles, died out before humans are thought to have arrived on the scene. According to this theory, it was drought that killed off the giant kangaroo.

In 2009, however, an analysis of the teeth of giant kangaroos revealed traces of drought-resistant plants, which was taken by some to point the finger back at humanity, as the animal had evidently adapted to climate change.

32. Why does Australia have so many venomous animals?

It is said that seven of the world's ten most venomous snakes are to be found in Australia. Fortunately, the snakes tend to avoid people and there has not been a death from snakebite in Australia for many years. On the other hand, box jellyfish, stonefish, and both funnel web and redback spiders *do* continue to kill people. All of which raises the question as to why so many species have evolved a deadly weapon against humans in a continent in which the human population has always been very sparse.

In the seas around Australia, the fatal attractions include the long tentacles of the box jellyfish whose powerful venom may cause excruciating pain and even death from cardiac arrest. The blue-ring octopus is another of the world's most toxic sea creatures: although only the size of a golf ball, it delivers a venom that paralyses its victim, with no known antidote. Perhaps most painful of all, however, is the stonefish, which lurks at the bottom of reefs, disguised as a rock.

Back on land, apart from the snakes, Australia offers the funnel web and redback spiders, of which the latter are known to have developed an unpleasant habit of nesting under lavatory seats. Thanks to the development of anti-venoms, deaths from spider bites are now very rare, but that does not alter their level of toxicity.

Snakes in India or scorpions in Mexico may be responsible for far more human fatalities, but the wide variety of fauna with anti-human capabilities in Australia is remarkable.

BATS

33. What is it like to be a bat?

In 1974 the American philosopher Thomas Nagel wrote a paper with this question as its title, and his essay has since become one of the most widely cited papers in any discussion of consciousness. Nagel argues that mental activity cannot be explained in terms of a physical process without losing the subjective experience. There can never be an objective account of a conscious experience. Or, to put it another way, only a bat can know what it is like to be a bat.

BEES

34. Why have half the honeybee colonies in the USA and Europe collapsed since 2006?

In Europe, it is known as honey-bee-colony depopulation syndrome, while in America they call it colony collapse disorder. Whatever the name, the result is the same: previously thriving colonies of honeybees can suffer catastrophic collapse. Since 2006, Europe and the USA have lost around half of their honeybees, and nobody quite knows why. Mites, parasites, fungi, pesticides or viruses could be to blame; even GM crops have been accused, and recent research has identified a parasite and a fungus that appear to have been present in all collapsed colonies; but the precise cause is still unknown.

35. How do bumblebees manage to fly?

Until 1996, bumblebees posed a big problem to the science of aeronautics, The problem was raised at the University of Göttingen in Germany in the 1930s, when a calculation was made that showed that according to everything that was known at the time, there was no way a bumblebee's wings, flapping at the rate they do, could possibly produce enough lift to enable the bumblebee to fly. Its body weight was simply too high to be kept airborne.

In 1996 researchers in Cambridge seemed to have found the solution. By building a model of a flying insect and analysing the forces acting on it, they discovered a previously unknown source of lift, created by vortices of air trapped around the creature's body. For some years, this allowed bumblebees to buzz around in peace, in the knowledge that their flight was scientifically possible after all. In 2001, however, Michael Dickinson and James Birch of the University of California came up with a more detailed picture of air flow over an insect wing, and in so doing cast renewed doubt on the possibility of bumblebee flight. After creating a robotic fruit fly that was more sophisticated than the Cambridge bumblebee, they concluded that the vortices identified in the earlier work could not explain the mystery of bumblebee flight after all.

Ever more complex models of flying insects followed, but there is still a discrepancy between theory and practice, possibly due in part to a difficulty in accurately simulating the rotation of an insect's wings during flight. A recent study compared the actual ability of bumblebees to lift weights with theoretical predictions of how much they could carry. While the latest models stated that a bumblebee should be able to lift its own body plus an additional 53 per cent of its own weight, the experiments showed that the weight-lifting abilities of bees are 18 per cent better than predicted. So there are clearly some aspects of a bee's flying ability that we still do not properly understand.

36. Do bumblebees have personalities?

The question of whether animals have personalities has been intriguing a number of researchers in recent years, and a flurry of papers have reported that creatures such as spiders, squid, blue tits and social bees have all shown behaviour indicating that individuals possess something analogous to human personality. For the purpose of these experiments, 'personality' is equated with 'individual-specific consistency in their behaviour across time and context'. In other words, if an animal shows an identifiably different behaviour to another of its species in response to a similar situation, and that difference is maintained over time, then the animal has a personality.

In 2010, researchers at London University reported the results of experiments to monitor the reactions of bumblebees when they encountered flowers of a colour they had not previously seen. Using artificial flowers with sucrose solutions at their centres, the researchers measured the time bees spent foraging at each flower. As is generally the case with animals encountering something new, they spent longer investigating the strangely coloured flowers, either out of interest (neophilia) or suspicion (neophobia), but the overall results fell short of confirming that bumblebees have personalities. In that respect, the experiments started well by showing that individual bees showed differing behaviours towards the new plants, but those differences were not exhibited consistently over an extended period: 'We conclude that for the neophilia/neophobia paradigm used here, bumblebee foragers do not fulfil the criteria for animal personality in the common sense of the term. Instead their behavioural response to novelty appears to be plastic, varying on a day to day basis.'

More research is clearly needed.

37. What information do bees obtain from watching the waggle dance of others, and how do they obtain it?

Ever since Karl von Frisch began to decipher the dance language of honeybees – for which he received a Nobel prize in 1973 – work has continued to understand this means of communication. We now know that the choreography of a bee's 'waggle dance' contains information about the direction, distance and quality of a food source. The dance starts with a run forwards, during which the bee conveying the information waggles its posterior and buzzes, then it turns right, returns along a semicircular path to its starting place, runs forwards again, then turns left and again returns in a semicircle. This figure-of-eight manoeuvre is repeated many times.

The direction of the initial forward run indicates the angle between the Sun and the food source, while the speed of the dance indicates its distance. Thanks to recent research, we know that European bees can communicate with Asiatic bees of the same species, though the Asiatic bees tend to get distances wrong until they get used to their visitors. We also know that when deprived of sleep, bees can still communicate distance accurately, but their indications of direction become unreliable.

What we do not know is how this information (and how much of it) is picked up by the other bees that are present when the dance is being performed. Besides the choreography, the dancing bee also buzzes and emits chemicals, which may be smelt by other bees. Experiments with robotic bees seem to suggest that the dance is enough, but other experiments seem to confirm that sound and chemistry also play a part. Doubt has also been cast on how accurately the audience perceive the intentions of the dancer, or whether there is still a strong hit-and-miss element in their attempts to follow instructions.

BIOLOGY

38. What is life?

Poets, theologians and scientists have all pondered this question at length, but none has come up with a totally convincing answer. From a scientific point of view, the question is how atoms and molecules of hydrogen, carbon, oxygen and other elements combine to form living plants, microbes, animals, human beings and everything else in the living world, which utilize energy and energy-giving items in their environment to grow and reproduce, and which after death turn back into inorganic material from which new generations may be formed. What is the literally vital (from the Latin *vita*, 'life') element that gives something life? Or, as the great Austrian physicist Erwin Schrödinger put it in his 1944 book *What is Life*, 'How can the events in space and time which take place within the spatial boundary of a living organism be accounted for by physics and chemistry?'

The subsidiary question, known as Schrödinger's paradox, is how life apparently manages to circumvent the second law of thermodynamics. According to that law, all closed systems – whether we are talking about the motion of the molecules of a gas in a test tube, or the collisions between balls on a snooker table, or the behaviour of the entire solar system – approach a state of maximum disorder. Without any external input, the energy within a system will tend to diffuse and disperse, causing all apparent order and organization to decrease. By contrast, life can only function if chemical elements form themselves into ordered structures that maintain their integrity. A living organism obtains energy from food or light and utilizes that energy to maintain a highly ordered state. Organisms are organized. Living organisms may decay and die just as inorganic materials corrode and crumble; the difference is that a living organism may leave behind the seeds from which a new generation will emerge. Once life appears, it seems to have a tendency to spread, bringing order rather than the increasing disorder predicted by the second law of thermodynamics. The only answer must be that life is not a closed system. The increase of order inside an organism must be more than balanced by an increase in disorder in the universe as a whole.

39. What causes ageing?

The search for an elixir of life that would negate or reverse the ageing process has occupied philosophers, alchemists and snake-oil salesmen for centuries, but there cannot be much hope of finding one until we know why organisms grow old and die. Since the 1980s, there has been a breakthrough in the understanding of the biology of ageing involving research into the function and operation of the strings at the ends of DNA molecules, called telomeres. With each reproduction of a cell, the telomere string has been found to grow shorter, thus acting as a sort of counter for the number of times the cell has reproduced. When the telomere has shortened to nothing, the cell stops reproducing and dies.

The optimists might say that all we need to achieve eternal youth is to find a way of resetting the telomere counter, or modifying the DNA in a manner that would stop the telomere-shortening process. An enzyme called telomerase is known to have such a function, and a good deal of research is being conducted to try to discover a method of turning that to youth-giving advantage. The question remains as to how telomeres evolved in the first place. One can only wonder whether, before telomeres appeared on the scene, all living organisms were immortal, unless some external event brought their lives to an end.

40. What causes a living cell to die?

Besides the telomere-related ageing process, cells die for two main reasons: necrosis (corresponding to illness or injury) or aptosis (an inevitable, biologically pre-programmed cell death). Aptosis, which has been described as a 'suicide mechanism', occurs as a response to signals that may come from outside (extracellular) or inside the cell itself (intracellular), and each of these types of signal may take many forms.

What links all the different reasons for cell death, and why cells need to undergo a constant process of reproduction and death anyway, are still unknown.

41. How do different cells know where to go during the development of an embryo?

A human body is made of around 100 trillion cells of about two hundred different types. All this begins with the fertilization of a single egg cell, which then divides and continues dividing, leading to the creation of innumerable cells. Our DNA may include the instructions for undifferentiated cells to turn into the right types in the right proportions to build the human body, but how do those cells, once formed, know where to go? What tells the cells designed to form the feet to head for one end of the building site, while the brain cells go to the other end?

Current research on this question concentrates on substances called morphogens, which influence both cell differentiation and cell position in the embryo. Some morphogens have been identified, but knowing *what* is responsible for giving cells their sense of direction is not the same as knowing *how* this information is put into operation.

42. How does the body regulate blood supply to its cells?

Cells need a blood supply in order to survive. The blood carries nutrients to the cells and carries away waste products, not directly but by filtering into the so-called interstitial space between blood capillaries and cells. The regulation of blood vessels is affected by a protein called VEGF (vascular endothelial growth factor), which comes in many different forms, some pro-angiogenic (encouraging the growth of blood cells) some anti-angiogenic (inhibiting it). If we could fully understand how VEGF works, that might supply us with a way of turning off the blood supply to cancerous tumours and starving them of the nutrients they need to survive.

43. Why do cancerous tumour cells migrate to different parts of the body?

Cancer results from a single cell that is genetically damaged and as a consequence undergoes a process of uncontrolled division and growth. Not only does it not know when to stop growing, but parts of the original cancerous growth, or primary tumour, may break off, travel through various routes to different parts of the body, and modify themselves into forms that can grow in the new environment as secondary tumours. If we knew the cause and mechanics of this modification process, we might be able to prevent secondary tumours from forming.

Beware of false knowledge; it is more dangerous than ignorance.

George Bernard Shaw (1856–1950)

44. Why can some creatures, such as salamanders, regrow lost limbs while others, such as humans, cannot?

Salamanders, flatworms and a number of other creatures can easily regrow lost body parts, including organs, muscle and nerves. Human regenerative capabilities, on the other hand, are much more limited: we can grow new skin and nerves, but an entire arm or leg is out of the question. Yet we know the body must possess the information needed to grow limbs, or it could not have done so during the embryonic stage.

Research on creatures that can regrow limbs, together with research on embryonic development, has identified the proteins responsible, as well as some of the genes that turn them on and off. One theory is that the key to regeneration lies in reproducing the conditions in the amniotic fluid during the development of an embryo. Indeed, some regeneration has been reported in mice with missing limbs that have been fitted with sleeves containing the right ingredients. Another theory holds that at an early stage of evolution, we all possessed the ability to regrow lost limbs, but while salamanders and worms have kept it, humans and other mammals have lost this ability, possibly through the development of a sophisticated immune system that prevents it from operating. The key to regeneration, if this is the case, would lie in unblocking the mechanism that stops this regenerative ability from functioning after the body is first formed.

45. Is there any hope that the people who have had their bodies – or just heads – cryogenically frozen may one day live again?

Cryonics is the procedure whereby a person's body (or just their head or brain) is frozen at death, the person concerned hoping that in the future they may be revived and cured. The number of people – mostly, if not all, Americans – who have undergone this procedure is estimated to be between 75 and 200. These figures include some who defrosted and had to be buried after the companies that froze them failed financially.

But is there any hope that such a procedure might work? Critics of cryonics, when it was first introduced in the early 1960s, very reasonably objected that the procedure involved the formation of ice crystals, which would cause irreparable damage to tissues, particularly in the brain. To prevent such injuries, the freezing technique was later improved by pumping the body full of cryoprotectants – chemicals such as the natural antifreeze found in various Arctic and Antarctic insects, fish, amphibians and reptiles.

On the general topic of resuscitation, there are two main schools or thought:

On the one hand, there is the argument that when you're dead you're dead, and nothing can change that. Since US law only permits the freezing of a body after death has occurred, that would rule out any chance of resuscitation.

On the other hand, sixty-one scientists signed an open letter in 2010 saying that in their view 'Cryonics is a legitimate science-based endeavour' and that 'there is a credible possibility that cryonics performed under the best conditions achievable today can preserve sufficient neurological information to permit eventual restoration of a person to full health'.

That 'credible possibility' could, of course, just mean that we know so little now about how the brain stores its information and memories that we don't know how much is irrevocably destroyed by death and the freezing process. If we do not know how something works or what has stopped it working, there is a credible possibility that we might one day find out and be able to mend it.

See also DNA 132–4, EVOLUTION 182–6

For lust of knowing what should not be known,
We take the golden road to Samarkand.

James Elroy Flecker (1884–1915)

BIRDS

46. Why do so many birds fly into windows?

According to Daniel Klem, who is probably the world's greatest authority on birds flying into windows, at least 225 species of birds have been seen flying into windows in the USA and Canada, and his estimate for the number of birds killed by flying into glass is somewhere between 100 million and 1 billion a year. Some say that birds fly into windows because they see the reflections of trees and grass in the glass; others say they are attacking their own reflection, which they see as another bird that needs to be chased from their territory. But however you (or the bird) looks at it, such birds clearly do not have the ability to perceive or understand reflections. In many ways, a bird's eye is a far more sophisticated and intricate organ than a human eye, yet we have no problem with glass or mirrors. A reflection-detection gene would have great survival value in the bird world, but as glass windows have only been in widespread use for a few hundred years, perhaps evolution has not had time to catch up. On the other hand, hedgehogs in the north of England have been observed scurrying across motorways instead of curling up in front of motor vehicles and being squashed, so life-saving evolutionary changes do sometimes happen relatively quickly.

We may, however, be on the verge of some interesting discoveries about the visual abilities of birds, thanks to research into our next unknown...

47. Do migrating birds use the Earth's magnetic field to navigate?

The question of how migrating birds find their way has long been a puzzle, but the idea that the Earth's magnetic field plays a part has been around for a long time. Only recently, however, has an explanation been offered as to how this might work. The key is the discovery of proteins called cryptochromes, which have been found to be sensitive to the extremely weak variations in the Earth's magnetic field. The theory is that if birds' eyes contained cryptochromes, they could 'see' the Earth's magnetic field and steer by it. As long as they do not fly into any windows on the way.

48. How much do chickens communicate with each other by clucking?

Researchers in Australia have claimed that chickens can convey at least twenty types of message in their clucking behaviour. Not only that, but their communication is sophisticated and involves high-level decision-making. Using an animated CGI rooster and monitoring hens' responses to its clucks and movements, Chris Evans and K-lynn Smith of Macquarie University, New South Wales, drew the conclusion in 2009 that chickens can effectively talk to each other. Not only do they have different clucks to warn other chickens of different types of predator, such as hawk or fox, but their clucking also gives information about the quality of food they have found. Most intriguing of all, it was found that roosters alter their food clucks according to who is listening. If there is a hen nearby, they squawk about the food as part of a courtship display, but if a larger male is in the vicinity who might steal the food, they keep quiet and rely on gestures to lure the hen to the dinner table. However, the limitations of what chickens can say, and the matter of whether it can be called language, are still open questions.

49. Why do birds interrupt each other's songs?

There are three main theories about why birds sometimes start singing before another bird has finished its own song. Some researchers have argued that it's an aggressive signal; some have said that it's a signal, but not necessarily aggressive; while others maintain that it just happens and doesn't mean anything at all. Whether a bird is interrupting another or merely joining in a duet may, of course, be difficult to determine. It is known, however, that zebra finches only sing duets with each other after they have become a couple.

See also DODOS 135, PENGUINS 359, SEX 417

BLACK HOLES

50. What happens at the centre of a black hole?

In 1931 the Indian physicist Subrahmanyan Chandrasekhar calculated that when a high-mass star collapses, the gravitational pull of its mass can be enough to compact it into an ever smaller space until it has shrunk to nothing, but a nothing of huge mass. In other words, it has a radius of zero and infinite density. Furthermore, its gravitational pull will be so great that anything – even light itself – within a certain distance (later called the event horizon) will be sucked in and never escape.

The whole concept seemed so preposterous to some astronomers that they refused to accept it, but later observations confirmed Chandrasekhar's theory, and his infinitely tiny yet infinitely mighty masses were dubbed 'black holes' (despite long resistance by the French, as the term, when translated into their language, means something rude).

Black holes are now accepted as part of the cosmological landscape, but the infinite density at their centres continues to pose problems. According to Einstein's theory of general relativity ($\rightarrow$ EINSTEIN 165), at the centre of a black hole space-time curvature becomes infinite and the pull of gravity is infinitely strong. Mathematically, the centre of a black hole is a singularity where space and time break down – as do the laws of physics themselves.

51. Which came first: black holes or galaxies?

Black holes have an important role to play in our theory of the formation of the universe. As more black holes were discovered, it was noticed that many galaxies had one at their centre – raising the question of whether the black hole had played a role in the formation of the galaxy itself. The standard picture once accepted by most astrophysicists had the Big Bang creating vast amounts of gases and energy; the gases then coalesced into the solid matter from which stars were formed; the stars then arranged themselves in galaxies through gravitational pull; and black holes were formed by stars burning out and collapsing. But recently a new possibility has been raised: was it in fact the black holes that had provided the gravitational pull to attract the stars and keep them in their galactic formation?

In 2009 astronomers in California estimated the mass of black holes in galaxies 12 billion light years away. Comparing the mass of a black hole with the total mass of its galaxy revealed a significantly higher figure than had been obtained for closer galaxies. Since the measurements, because of the distance, relate to a situation that pertained 12 billion years ago, and black holes cannot diminish in size, the Californian team concluded that the other galaxies must have grown around the black holes at their centres. As Christopher

Carilli of the National Radio Astronomy Observatory put it: 'Black holes came first and somehow – we don't know how – grew the galaxy around them.' The study, however, only took in four galaxies, and other astronomers have suggested that they may not be typical and that no clear conclusions can be drawn.

52. Why is the Sun moving so fast?

We all know that the Moon orbits the Earth about once a month and that the Earth orbits the Sun once a year, but it is easy to forget that the Sun is itself orbiting the centre of the Milky Way galaxy, moving at a speed of about 220 km (135 miles) per second in an orbit that takes about 240 million years to complete. The problem is that according to everything we know about the Milky Way and the laws of planetary motion devised by the German mathematician Johannes Kepler (1571–1630), the speed of the Sun ought to be only 160 km (100 miles) per second.

The Sun is not the only star moving at a speed different from that predicted. Accounting for such discrepancies, both in the Milky Way and other galaxies, is one of the major problems in cosmology.

Observations from the Hubble telescope have suggested a strong relationship between the rotational speed of stars and the mass of the black hole at the centre of their galaxy, but the reasons for this are unknown. A disc of dark matter (→ COSMOLOGY 107) at the edge of the galaxy could explain it, but another possible explanation involves a change in the laws of physics for objects on a galactic scale.

53. How many black holes are there in the Milky Way?

Until the early years of the present century, the existence of a black hole at the centre of the Milky Way was a matter for speculation. But as the evidence of its gravitational influence grew, the presence of a super-massive black hole at the heart of the Milky Way became undeniable. In recent years it has been located in the constellation of Sagittarius, and its mass estimated at 2 or 3 million times that of our Sun.

In 2009 a team of Harvard astrophysicists researching the early universe suggested that there could be hundreds of black holes in the Milky Way, the result of collisions with other galaxies a long time in the past. The Harvard team said that as their theory was a new one, nobody had been looking for such objects – and without the usual shining star cluster around a black hole, their proposed black holes would be 'all but impossible to find'.

BOUDICCA

54. Where did Boudicca fight her last battle with the Romans?

There is an oft-repeated story dating back to the 1930s that Boudicca, the queen of the Iceni who rose up against the Roman conquerors of Britain, is buried underneath a platform at King's Cross Station in London (though different accounts give different platform numbers). Whether this story was originally a hoax or based on a misunderstanding of a place name is unknown. The site of King's Cross used to be the village of Battle Bridge, but the word 'battle' in that name is thought to be a corruption of 'broad ford' rather than referring to any specific battle.

Leicestershire, Warwickshire, Essex and Northamptonshire have been suggested as possible locations of Boudicca's final defeat by the Romans, in AD 60 or 61. But all these surmises appear to be based not on any hard evidence for a particular location, but rather on the assumption that her army would have retreated north from London by the Roman road known as Watling Street.

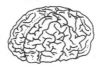

THE BRAIN

Let's admit it: we just don't know how brains work. Ever since thinkers started pondering the question, there has been a tendency to adopt the latest technological ideas to try to explain mental processes. The ancient Greeks saw the brain as an elaborately engineered plumbing system; much later, in the early twentieth century, it became an electrical circuit; and most recently the brain has been compared with a highly complex computer network. Every theory seemed to explain some aspects of brain function while ignoring others, yet ever since the seventeenth-century French philosopher René Descartes, and perhaps for even longer, one question has remained a matter of intense debate among philosophers and every type of neuroscientist:

55. How are the mind and the brain connected, if at all?

The brain, as we see it, is an organ for receiving and processing information. We can now look at individual neurons (nerve cells) in the brain and watch them working. Thanks to highly sophisticated brain-scanning methods, we can watch a brain as its owner thinks and see which parts are active. To some extent, we can therefore discern the mechanics of mental processes. But while we are thinking, we may know what we are thinking about; we have subjective experiences connected to those thoughts; we may even decide what we want to think about. This subjective side of our mental activity is called the mind (which is closely related to, though not identical with, the concept of consciousness). How we see the mind–brain relationship essentially comes down to two (or possibly three) choices:

(i) All events in the brain operate in accordance with the same laws of physics, so the brain completely determines all our behaviour, including subjective experiences. The mind, in other words, is just part of the brain.

(ii) The mind is a higher level process than the brain. It may both influence and be influenced by what goes on in the brain, but is a radically different type of process.

(iii) Subjectivity is just an illusion, so there's no mind–brain problem anyway.

This problem cannot be resolved until we know a great deal more about how the brain functions and what consciousness really is.

56. Will it ever be possible to read someone else's thoughts by examining their brain?

Shortly after the discovery in the early 1950s of the DNA double helix and the associated information-carrying messenger-RNA molecules, there was hope in some quarters that this would lead to an understanding of how the brain functions. In particular, some scientists thought that memories or new concepts or any other new piece of information corresponded with a newly formed RNA molecule in the soup of brain chemicals. If this was right, it should be possible, the theory went, to extract information from one animal's brain and inject it into another animal of the same species. Several experiments – with fish, worms, rats and other creatures – were performed to test the theory. Sadly, before any strong conclusions could be drawn, our understanding of the way RNA works made it clear that codifying memories could not be part of its function and the memory-transfer experiments stopped.

Quite how knowledge resides in the brain is still an open question. Individual neurons may be stimulated to produce specific mental responses, but it could be, depending on one's answer to the previous question, that what we think of as knowledge is just the mind's way of interpreting a neuron and the links called synaptic pathways that lead from one neuron to another.

Perhaps one day a complete map of the 100 billion or so neurons plus the trillions of synapses in an individual's brain will tell us everything we need to know about what they are thinking. But for the time being at least, we can't tell what someone is thinking by examining their brain activity. But if we cannot read what a brain is thinking by monitoring it with scientific equipment, we might ask whether it is possible through some sort of direct brain-to-brain exchange of information – which bring us to the next question:

57. Is telepathy possible?

Since the nineteenth century, a vast number of experiments have been conducted to test whether mind-reading is a reality. Rationalists point to the many hoaxes and frauds in this area, and to the fact that, even when an apparently well-designed experiment has given a result significantly supporting the telepathy hypothesis, such experiments have always proved impossible to repeat reliably. Parapsychologists, on the other hand, tend to assert that while no individual proof of telepathy has ever been demonstrated, the large number of published studies supporting it taken as a whole cannot be explained away as mere chance or, indeed, by any means other than accepting that some degree of telepathy has taken place. During the Cold War years, both the Soviet and US military intelligence organizations believed so much in the possibility of telepathy that they expended a great deal of energy in experiments in 'remote viewing' (not to mention trying to kill goats by staring at them), but the results were, to say the least, unconvincing.

58. Why have human brains been getting smaller for the past 20,000 years?

The first human beings of the genus *Homo* evolved over 2 million years ago. *Homo sapiens* has been around for some 200,000 years. Examination of skull sizes in the fossil record show that for nearly all of human history our skull sizes have been increasing. That appears to make sense. Bigger brains, at least in relation to body size, seem to indicate greater processing capacity and a higher level of intelligence. Yet in the past 20,000 years, comparisons of cranial capacity show that our brains have been getting smaller. This is difficult to explain, though several theories have been advanced, including the following:

(i) Our brains may have got smaller overall, but certain parts of the brain, such as the cerebellum (which controls motor movement coordination), have grown. The result is a smaller but more efficient brain.

(ii) Larger brains may not always be a sign of intelligence. Wolves have larger brains than dogs and are better at problem-solving, while dogs are better at carrying out learned tasks. Early humans may have had more problems to solve.

(iii) We are getting stupider.

Further research is clearly needed.

59. Why do people with more friends have larger amygdala?

The amygdala, two almond-shaped portions of the brain deep in the temporal lobes, is the part of the brain responsible for emotions. The size of the amygdala has been associated, both in humans and other primate species, with the size of an individual's social circle. Indeed, in a paper published at the end of 2010, it was reported that amygdala size correlates with the number of 'friends' people have on Facebook.

Whether increased social activity leads to growth of the amygdala or a large amygdala encourages social activity is unclear. There is also the question to what extent the number of a person's Facebook 'friends' tallies with their number of real friends, and also the question of how either of these numbers relates to the intensity of their social calendar.

60. How can a single brain cell hold the concept 'Brad Pitt and Jennifer Aniston', while remaining unresponsive to each of the individuals involved?

In 2005 Rodrigo Quian Quiroga of the California Institute of Technology in Pasadena published research that has raised all sorts of questions about how the brain works. Studying epileptic patients who had been implanted with devices to monitor brain-cell activity, he was able to detect individual neurons that became active in response to the patient seeing pictures of well-known people. In one patient, he identified a cell that only responded to pictures of Bill Clinton. Another had a Jennifer Aniston cell, and so on. Most remarkably, however, he found one case of a neuron that responded to pictures of Brad Pitt and Jennifer Aniston together, but remained inactive when the patient was looking at either Brad Pitt or Jennifer Aniston alone.

The idea of a 'grandmother cell' was coined in 1959 by the neurobiologist Jerome Lettvin to mock the suggestion that a single cell could be responsible for identifying an individual person. Nobody now believes that the elimination of one neuron could cause us to forget dear old Granny, but Quiroga's research clearly demonstrates hitherto unsuspected complexity in what information a single cell may hold. However, whether the Pitt + Aniston cell can be reprogrammed to become a Pitt + Jolie cell, as the leading man changes partners, is unknown.

61. Are sex and violence linked in human brains as they seem to be in mice?

According to a report published in 2011, experiments on mice have identified a small cluster of brain cells that come to life when the mouse is fighting, and also when it is having sex. When these cells are stimulated in a male mouse, it will attack any other male that comes near – even if the other male has been castrated or anaesthetized, which is normally enough to make another male mouse ignore it. The stimulated mouse will also attack females, but only if he is not already mating with them. On one occasion the male concerned even attacked a stuffed laboratory glove.

The results seemed to show that sex and violence are linked in the mouse's brain. 'I think there's every reason to think that this would be true in humans,' said David Anderson, one of the researchers responsible for the finding. As someone once asked, 'Are you a man or a mouse?' Perhaps, as far as sex and violence are concerned, it could be too close to call.

62. How do general anaesthetics work?

Well, they consist of a cocktail of drugs that put you to sleep, relax your muscles and prevent you from feeling anything – and you don't remember anything, unless something goes terribly wrong. We all know that, but quite how they do the job so well is still not completely understood. The general view is that general anaesthetics operate directly on the central nervous system to inhibit synaptic transmission, in other words, they interfere with the way neural impulses are transmitted between adjacent neurons. This results in a general loss of consciousness that affects sensory awareness in all forms and in all parts of the body. Yet the precise biochemistry behind the whole process has yet to be explained.

63. Are cell phones bad for us – or might they be good for brain function?

Ever since mobile phones became popular in the early 1990s, there have been concerns, and scare stories, about them being a health risk and even a possible cause of brain cancer. Research has failed to confirm these dangers, but in February 2011 a study was published claiming to show that spending 50 minutes with a cell phone at one's ear *does* change brain-cell activity.

Specifically, the researchers found that glucose metabolism, which is a general sign of brain activity, increases in the area of the brain closest to the phone's antenna. The significance of this, the researchers and other experts have said, has yet to be assessed. As we all know, spending fifty minutes sitting next to someone who has a cell phone at his ear can also be a stressful experience, but that is another matter.

See also CONSCIOUSNESS 101–5, LANGUAGE 254, MEMORY 288–95

BRUSSELS

64. What is the origin of the small boy depicted in the Mannekin Pis statue in Brussels?

One of the most popular tourist attractions in Belgium is the statue in Brussels known as *Mannekin Pis* ('little man pee'), depicting a naked little boy urinating into a fountain. The sculpture was designed by Jerome Duquesnoy and dates back to 1618 or 1619, though it replaced an earlier similar statue that may have dated back to the fourteenth century. Yet the reason for the little boy's quaint pose, in either the original or its replacement, are buried beneath piles of myth.

One story is that the statue celebrates an incident in which the little boy put out a fire by peeing on it, thus saving a king's castle from burning down. A similar tale has the little boy urinating on the fuse of explosives left at the city walls by an invading force.

Another version says the statue is of the infant Duke Godfrey III of Leuven, who is said to have urinated on enemy troops from a position high in a tree where he had been lifted in a basket to keep him safe from the battle.

Or you may prefer the tale of a rich merchant whose son went missing and who vowed to commission a statue of the boy exactly as he was found.

BUTTERFLIES

65. How do monarch butterflies know where to go on their long migrations?

The monarch butterfly is an extraordinary creature. Each year, beginning in August, vast numbers of them begin a migratory flight from Canada and the north of the USA to Mexico. The journey, which takes two to three months, covers between about 3,000 and 5,000 km (2,000–3,000 miles). It is possible that none of the butterflies that embark on the journey reach the destination: monarchs are prolific and fast breeders, and the butterflies that arrive in Mexico in the autumn may be three or more generations separated from those that began the migration at the end of summer.

As many as 300 million monarchs may eventually spend the winter in Mexico, of which around half will have died before the flight back in spring. Then once again the two- or three-month flight may involve three or more generations, before the monarchs reach their summer home. Unanswered questions remain: when they emerge from their chrysalises, how do the young butterflies know where they are on the migration route, and what cues do they pick up from their environment to show them the way ahead?

66. If you teach a caterpillar something, will the butterfly it turns into retain that knowledge?

Earthworms have been taught to turn one way or the other in a T-shaped enclosure. Just place food at one end or the other of the top bar of the T and the creature will learn. It is reasonable to assume that caterpillars would be similarly easy to teach. But what, if any, of this training would be carried over when the caterpillar metamorphoses into a butterfly? Does a left-turning caterpillar turn into a leftwards-flying butterfly?

As the main task of the caterpillar brain must be to keep its body working satisfactorily, it must be radically different from the butterfly brain, which has a totally different set of bodily priorities. But even though the caterpillar's brain turns to soup as it metamorphoses into a butterfly, it would be interesting to know if anything in the caterpillar's mental life is carried through to its role as a butterfly. No research seems to have been done in this intriguing field.

The sum part of ignorance that we arrange and classify we give the name knowledge.

Ambrose Bierce (1842–1913)

CANNIBALISM

67. Was cannibalism ever a normal human practice?

In William Arens's 1979 book *The Man-Eating Myth: Anthropology and Anthropophagy*, the author argues that throughout the ages people have spread tales of anthropophagy (i.e. cannibalism) in order to discredit their enemies and establish their own cultural superiority. He concluded that cannibalism was never a widespread practice, and that accounts of it are generally steeped in racism and are over-dependent on hearsay evidence. In the opposite corner are Daniel Diehl and Mark P. Donnelly, who in their 2006 work, *Eat Thy Neighbour: A History of Cannibalism*, insist that the practice was widespread in ancient times.

68. Did our Stone Age ancestors eat their own dead?

Almost all cultures include horror stories of flesh-eating ogres in their myths, but convincing evidence of widespread cannibalism among ancient peoples is very rare. In the 1990s a team of French and American archaeologists began investigating a 100,000-year-old Neanderthal site in a cave at Moula-Guercy close to the River Rhône in southern France. In 1999 they reported that they had found human bones from which the flesh had been removed in the same way that the Neanderthals butchered animal meat. This appeared to suggest cannibalism, but the archaeologists could not say whether this was part of a ritual, or an act of desperation at a time of famine, or whether it was standard practice among the Neanderthals of the time.

In 2009 a much larger collection of butchered human bones was found in the village of Herxheim in southwestern Germany. This appeared to present much clearer evidence of everyday cannibalism – and among modern humans, for these remains date from the early Neolithic period, between 7,000 and 7,500 years ago. Researchers concluded that over a period of a few decades, people at this site ate hundreds of their fellow humans.

Some archaeologists suggest that the evidence points not to cannibalism, but to a ritual practice in which bodies that had previously been buried were disinterred, dismembered and the flesh removed, before being reburied. To the sceptics, scratch marks on bones do not necessarily mean that our ancestors were cannibals.

69. Were the Anasazi people of the American Southwest cannibals?

In 2001 in New Orleans the Society of American Archaeology held a symposium under the title 'Multidisciplinary Approaches to Social Violence in the Prehispanic American Southwest'. Under the cloak of that academic title lay one simple question: Did the Anasazi, who were the ancestors of the Pueblo Indians of the southwestern USA, routinely eat each other, roasted or boiled, in the period between AD 900 and 1200?

There are two sides to the debate, dubbed 'the bleeding hearts' and 'the rip-their-hearts-out' factions by the Colorado archaeologist Steven Lekson. The pro-cannibalism faction cited the usual evidence – cut marks, abrasions and marks that human bones and cooking pots could have left on each other. The anti-cannibalism faction proposed various alternative theories, suggesting that the bodies may have been prepared by reburial, or may have been preyed upon by wild animals, or might even have been the corpses of executed witches. The matter may only ever be resolved if archaeologists discover either a recipe book from the period or a well-preserved body with human flesh in its stomach.

CARTOGRAPHY

70. How did Martin Waldseemüller know about the Pacific Ocean for his 1507 map of the world?

In 1507 the German cartographer Martin Waldseemüller (→ AMERICA 3) published his *Universalis cosmographia secundum Ptholomaei traditionem et Americi Vespucii aliorumque lustrationes* ('The universal cosmography according to the tradition of Ptolemy and the voyages of Amerigo Vespucci and others'), a map that charted the New World discoveries begun by Christopher Columbus. His map depicted not only the east coast of the Americas, as explored by the early navigators, but for the first time showed the hitherto unmapped ocean later to be known as the Pacific. Yet the Spanish explorers Ponce de León and Vasco Núñez de Balboa did not set eyes on the Pacific Ocean until 1512 or 1513, at least five years after Waldseemüller's map. There had, of course, been speculation that America was a new continent and not, as Columbus had thought, the other side of Asia, but some of the distances measured on the map are uncannily accurate. Either Waldseemüller knew something that we do not know he knew, or he made a very lucky guess.

71. How did the Piri Reis map of 1513 give such an accurate picture of Antarctica?

Discovered in an old library in Turkey in 1929, this map drawn on gazelle skin was the work of Piri Reis, a famous Ottoman admiral, geographer and cartographer of the sixteenth century. Piri Reis – more properly Haci Muhiddin Piri oglu Haci Mehmed

– acknowledged his debt to various extant maps – some even then a thousand years old – in the collection of the Imperial Library of Constantinople, to which his rank gave him access. Remarkably, the Piri Reis map shows not only the western coast of Africa and the eastern coast of South America, but also the northern coast of Antarctica – in perfect detail. Even more extraordinarily, he shows the coastline of the land beneath the ice, though geological evidence shows that the latest this could have been charted in an ice-free state is 4000 BC, if not even earlier. So who made the map from which Piri Reis obtained his information – and where and when can it have been made?

It's because someone knows *something* about it that we can't talk about physics. It's the things that nobody knows about that we can discuss. We can talk about the weather; we can talk about social problems; we can talk about psychology; we can talk about international finance... so it's the subject that nobody knows anything about that we can all talk about!

Richard Feynman (1918–88)

CATS

72. Why are female cats right-pawed and tom cats left-pawed?

In 2009, scientists at Queen's University, Belfast published the results of a study on paw preference in cats. When cats are playing with a fishing-rod toy, the scientists found that they were equally likely to use either left or right paw, but when posed with the more complex task of getting food from a glass jar, male cats were found to show a strong preference for using their left paw, while females used their right.

In humans, left-handedness has been associated with the hormone testosterone (which has been used to explain why more men than women are left-handed). Exposure to testosterone has also been shown to result in a female cat changing her paw preference from right to left. Scientists do not yet know why the hormone has this effect, or whether it is testosterone that is responsible for the initial handedness.

See also HANDEDNESS 222–4

73. What colour was Christopher Smart's cat Jeoffry?

While incarcerated in London's St Luke's Hospital for Lunatics between 1757 and 1763 with only his cat for company, the poet Christopher Smart wrote a poem of over 1,200 lines called *Jubilate Agno* ('Rejoice in the Lamb'), of which 74 lines are devoted to his pet. Beginning with the words 'For I will consider my Cat Jeoffry', he extols the cat's personal and religious virtues at great length.

After Smart was released from St Luke's, he quickly ran into debt and died in a debtors' prison. What became of the cat is not known, and he never told us what colour it was. More than one recent publication of the full poem, however, include illustrations depicting Jeoffry as a marmalade cat. This has no doubt been prompted by Smart's line 'For he is of the tribe of Tiger', although this might equally suggest a tabby. Smart may simply be comparing his small feline with the fearsome big cat of the forests of southern Asia, or even using 'tribe of Tiger' simply to mean the cat family – rather than telling us Jeoffry had black and orange stripes.

CHEMICAL ELEMENTS

74. Can untriseptium, aka the theoretical element 137, aka feynmanium, ever exist?

When in 1869 Dmitri Mendeleyev presented to the Russian Chemical Society his periodic table of the chemical elements, he showed how they could be listed by their atomic number (which turned out to be equal to the number of protons in their nucleus) and in groups that shared similar, recurring properties. His table had certain gaps, which he believed would be filled by the discovery in the future of hitherto unknown elements, whose properties he successfully predicted. Some ninety elements occur naturally; those of greater mass are highly unstable and radioactive, and can only be created for very short periods in the laboratory. In recent years elements have been synthesized with atomic numbers up to 118.

The great US physicist Richard Feynman, who died in 1988, once pointed out that according to a simple interpretation of theories proposed by Niels Bohr and Paul Dirac, no element with an atomic number greater than 137 can exist, because the mathematics would then give a speed for its electrons greater than that of light, which is impossible. For that reason element 137, or untriseptium, is also known as feynmanium. Attempts to synthesize it, however, have so far failed.

I can live with doubt, and uncertainty, and not knowing. I think it's much more interesting to live not knowing than to have answers which might be wrong.

Richard Feynman (1918–88)

75. Does californium exist naturally on earth?

As mentioned above, when in 1869 Dmitri Mendeleyev presented his periodic table of the chemical elements, he posited the existence of hitherto unknown elements and predicted what properties such elements would have. In many cases, these unknown elements tallied with unexplained lines in the solar spectrum (→ THE SUN 455), which gave some hints to researchers as to how they might synthesize such elements. As scientists began to understand the concept of radioactive decay – by which one atom could change into another – they developed methods of encouraging the formation of the missing items in Mendeleyev's table. One such item was the new element discovered at the University of California in 1950, and shortly afterwards named in honour both of that university and of the State of California.

Californium has atomic number 98, and has been found to have a number of applications, from the treatment of cancer to the start-up processes of nuclear reactors. Around twenty isotopes of californium have been discovered, all highly radioactive, but with half-lives (the average time they take to decay) varying from a few minutes to 898 years.

Since californium has been detected in the spectrum of the Sun, and it is known to be formed by the effect of nuclear radiation on other atoms, it may well have been present on Earth in the early days of the planet, but taking less than a thousand years to decay, it could not have survived since then. The question of whether any californium is still being produced naturally on Earth is still unanswered. It has been found in the radioactive dust after nuclear explosions, and it is thought that minute amounts may be produced by nuclear reactions in uranium ores. Such traces of californium as have been found so far, however, have all been near facilities that use synthetic californium for medical or prospecting purposes, so presumably were man-made.

76. How far can the periodic table extend beyond the 118 known elements?

Although the Dirac and Bohr equations suggest that no element with an atomic number greater than 137 can exist (→ 74), they do not take relativity into account. Some calculations have suggested that the elements end at atomic number 139, while others suggest they may theoretically extend up to 173. Of the 118 known elements, of which the latest, rather unimaginatively named ununoctium, or one-one-eight, was identified in 2002 and synthesized in 2006, only numbers 1 to 94 are known to occur naturally on earth. Of those, only about eighty are considered stable, while the others decay radioactively over timescales varying from minutes to billions of years. Even some of the elements generally considered stable are thought to decay eventually, though the process may take considerably longer than the universe has been in existence.

The trouble with the heavier elements is that the repulsion forces between the protons tend to overwhelm the strong nuclear force binding the atom together. For this reason, the atom, often created under extreme conditions of temperature or pressure, only exists for the tiniest fraction of a second. We do not know what possible limits there may be to determine whether it can exist at all.

CHIMPANZEES

77. Why do ambidextrous chimpanzees eat fewer termites?

A good deal of recent research has shown that many chimpanzees in the wild as well as in captivity show a preference for using one hand over the other. This has raised some curious anomalies. On the admittedly simplistic principle of two hands being better than one, one might expect ambidextrous chimps to be the most successful, and that indeed seems to be the case in some trials, yet when it comes to fishing out termites with a twig and eating them, ambidextrous chimps have been found to end up with smaller meals than their right-handed or left-handed colleagues.

One suggestion is that ambidextrous chimps end up with two hands that both perform normal tasks adequately, while the others have one hand that is rather weak or clumsy and the other that is very good, and which is therefore better suited to specialist tasks such as fishing for termites with a twig. Before we can claim to understand the problems of the ambidextrous chimp, a good deal more research is needed comparing the performances in a wide variety of tasks of ambidextrous and non-ambidextrous chimps.

78. Why don't chimpanzees ask questions?

The argument over whether non-human primates can develop language skills has raged for almost half a century, since researchers started to teach chimpanzees sign language. The most celebrated chimps in that respect were Washoe (1965–2007) and Nim Chimpsky (1973–2000), the latter being punningly named after the US philosopher Noam Chomsky, who wrote extensively on language. Each developed a vocabulary of over a hundred signs, which it was claimed they manipulated in ways that indicated genuine linguistic ability.

Some observers, however, maintained that all the chimps were doing was exhibiting conditioned responses and selecting the appropriate signs they had learned. These sceptics were not convinced this constituted evidence of thought being put into words.

While Washoe, Nim Chimpsky and other chimpanzees did seem able to understand variations in word order and to manipulate signs in a way that could be taken to indicate a form of communication, one curious linguistic ability has remained absent in our primate cousins: they do not seem to be able to ask questions. Why they have this particular linguistic block has not been explained.

See also DISEASE 126, DOGS 136

This gray spirit yearning in desire to follow knowledge like a sinking star beyond the utmost bound of human thought.

Alfred, Lord Tennyson (1809–92)

CHRISTIANITY

79. Was there ever a real Holy Grail?

The Holy Grail is a sacred but elusive object that figures in Christian and literary tradition. It is said to be the cup or bowl used by Jesus at the Last Supper, and is believed to have miraculous powers. In the Middle Ages the Holy Grail played an important part in the Arthurian cycle of legends, first being mentioned in a poem by Chrétien de Troyes entitled *Perceval, le Conte du Graal* (*Perceval, the Story of the Grail*), written between 1180 and 1191. Shortly afterwards, Robert de Boron gave the Grail added significance and mystery in his poem *Joseph d'Arimathe*, in which Joseph of Arimathea (who in the Gospels gives up his tomb for the burial of Jesus) is given the Grail by an apparition of Christ himself, and then uses it to collect drips of Christ's blood when he is taken down from the Cross.

As more than a millennium divides the Crucifixion and the first reference to the Grail, it looks very much like a myth, but several churches hold artefacts held to possess certain Grail-like characteristics. The most significant of these is the Holy Chalice in Valencia Cathedral, Spain, a vessel that has been the official papal chalice of many popes and which was used by Pope Benedict XVI during a mass held in the cathedral in 2006. It was given to the cathedral by King Alfonso V of Aragon in 1436, but was supposedly brought to Spain by St Lawrence in the third century. Some archaeologists have dated the chalice to the first century AD, and many Christian historians say that, of all the various artefacts claimed to be the Holy Grail, this particular vessel is most likely to be the genuine article.

80. How did the image of a man become imprinted on the Turin Shroud?

The Shroud of Turin is a much-revered piece of ancient linen cloth kept in Turin Cathedral, Italy. On the Shroud can be seen the image of a man, traditionally said to be an impression of the body of Christ after he had been taken down from the Cross. The Shroud has posed its mysteries for so long that there is even a word for its scientific study: sindonology (from the ancient Greek word *sindon*, meaning a burial cloth). Yet even the most diligent sindonologists would admit that certain aspects of this holy relic are baffling.

The Shroud has been well documented since the fourteenth century, though whether it dates from that period or truly is the cloth in which Christ's body was wrapped after the Crucifixion is still a matter of dispute. After intense debate, small samples of the cloth were made available for carbon-dating in 1988 and the results suggested, with a 95 per cent degree of certainty, that it dated from between 1260 and 1390. Subsequently, however, both the testing methods and the nature of the sample tested have been questioned. There is some evidence that the fabric sample may not have been typical of the original cloth as a whole but the result of later repair or tampering.

Whatever the date of the cloth, the origins of the image on it are a complete mystery. Signs of injury to the body imprinted on it are consistent with what is known of crucifixion methods at the time of Christ – but this knowledge would not have been available in the fourteenth century. The image also becomes clearer when viewed as a photographic negative, yet it dates from long before photography. Finally, no attempt to reproduce such an image on cloth has ever succeeded.

81. Was the patron saint of Ireland a Scot or a Welshman?

St Patrick is the patron saint of Ireland, but the only reliable details of his life consist of two letters known to have been written by him, and they are rather short on facts. He gives his place of birth as Banna Venta Berniae in Britain, which some maintain is in southwest Scotland, just over the border from Carlisle, and others have placed in Wales. He was kidnapped by pirates and taken to Ireland at the age of sixteen, but subsequently escaped and returned home, before going back to Ireland to spread the word of Christ. He died in AD 493 on 17 March, according to the old Irish annals, but even that is not entirely certain, as until recently it was generally believed that St Patrick had died in 420.

There are similar gaps in our knowledge of the identities of many popular saints, including St George and St Valentine.

82. Did the early Christians have a hand in writing the Dead Sea Scrolls?

In the winter of 1946–7, a Bedouin shepherd made an astonishing discovery when he accidentally fell into a cave near the ruins of Khirbet Qumran on the northwest shore of the Dead Sea. As he clambered out, he brought with him a handful of ancient scrolls. Over the next few years, many more scrolls were recovered from this and a number of other nearby caves. They were found to date from between 150 BC and AD 70, and contained 972 texts from, and commentaries on, the Jewish Bible, written in Hebrew, Aramaic and Greek. By far the oldest original Biblical texts, the Dead Sea Scrolls immediately attained a religious and mystical significance that added to the debate over their origins, translations and relevance.

The most natural conclusion concerning their origins was that they were sacred texts of the Essenes, a devout Jewish sect of the period whom Pliny had said lived on the west side of the Dead Sea. Since 1990, however, alternative theories of the scrolls' origins have been proposed. Further archaeological evidence from Qumran, together with analysis of the content of the scrolls, has suggested to some that they may have been written not by the Essenes, but by another Jewish sect in Qumran; or they may have been compiled in Jerusalem, then later taken to the Qumran caves for safe-keeping. Other scholars have taken a fragment of St Mark's Gospel in the scrolls as evidence that some of the texts may have been written by early Christians.

83. Who wrote the New Testament Gospels?

Nobody is sure when the Gospels of Matthew, Mark, Luke and John were written, or by whom, but most agree that the Gospel of Mark was the first to be written, around AD 65; Mathew and Luke followed shortly after, perhaps between AD 65 and 70; while John is thought to be the last, dating from around AD 100. These conclusions are largely derived from analysis of the texts, with both Matthew and Luke quoting extensively from Mark, while John recounts different events in a different style. But who were Matthew, Mark, Luke and John?

The author of Matthew is traditionally identified with the tax collector mentioned in Matthew 9:9 (also known as Levi), but if that is the case, and he was so strongly connected with Jesus, why does he rely so heavily on the writings of Mark?

Mark himself was not an apostle but an associate of the apostle Paul for a short time, though some authorities insist that his Gospel is more closely based on the preaching of Peter.

The Gospel of Luke is thought to have been written by the same person as the Book of Acts, for one follows the other in a natural way, and both are addressed to a Roman named Theophilus. Apart from that, all we seem to know about Mark is that he was a doctor.

The authorship of St John's Gospel is the most mysterious of all, and the text never even mention's John's name. The reference to 'the disciple that Jesus loved' in John 13:23 is generally taken to refer to the author of the Gospel, but that doesn't help us identify him. It is generally accepted, however, that he wasn't the same John who wrote the Book of Revelation. *That* John had his visions of the Apocalypse on the Greek island of Patmos, and is known as St John the Divine.

See also JESUS CHRIST 249–51

CLEOPATRA

84. What did Cleopatra look like?

The Greek historian Plutarch, in describing the great Egyptian queen who lured both Julius Caesar and Mark Antony into her bed, wrote that 'her beauty... was in itself not altogether incomparable, nor such as to strike those who saw her; but her conversation had an irresistible charm'. The Roman historian Cassius Dio, on the other hand, says: 'She was a woman of surpassing beauty.'

On a more specific point, the seventeenth-century French philosopher and mathematician Blaise Pascal, in his *Pensées*, wrote: 'Cleopatra's nose, had it been shorter, the whole face of the world would have been changed.' We should remember, however, that at the time, large noses were seen as a symbol of strength. Sadly we are unable to check this, as most of the statues of Cleopatra have had their noses knocked off. In any case, portrayals of Cleopatra in statues and on coins are highly inconsistent. According to a study at Nottingham University in 2007, based on her face on a coin, Cleopatra was a pointy-nosed, thin-lipped woman with a jutting jaw.

85. Did Cleopatra really die of a snake bite, or was she poisoned?

According to the Greek historian and philosopher Strabo, who lived at the same time as Cleopatra, Cleopatra took her own life either by applying a poisonous ointment or by allowing herself to be bitten by an asp, a small, venomous snake. Later Roman writers chose the asp version, some even upping it to two asps. They all agreed that she was bitten on the arm, but Shakespeare added a bit of drama by having her clasp the asp to her breast.

Other historians, however, have suggested that this is all romantic nonsense and that her Roman vanquisher Octavius (later the Emperor Augustus) had her killed. The latest diagnosis, offered by the German historian Christoph Schaefer in 2010, concludes that a snake could not have caused the slow and pain-free death reported, and that the fatal cocktail she took most likely consisted of a mixture of hemlock, wolfsbane and opium.

> Woe unto you, lawyers! For ye have taken away the key of knowledge.
>
> The Gospel of Luke 11:52

86. Where are Cleopatra and Mark Antony buried?

The final resting places of both Cleopatra and her lover Mark Antony are unknown. According to Plutarch, Octavius allowed them to be buried together, but there are no known accounts of where the burial took place. Various sites around Alexandria have offered hope to archaeologists that the tomb may be found, but excavations have so far failed to reveal the secret.

CLIMATE

87. How much of the current climate change is due to greenhouse gases?

In the 1970s, following a number of very cold winters, there was a flurry of panic suggesting the approach of a new ice age and blaming it on greenhouse gases. When these gases accumulate in our atmosphere, they have two effects. One is a warming effect, in which they act like a duvet and stop the Earth's radiated heat from escaping into the atmosphere. The other is a cooling effect, in which the gases act as a barrier to the Sun's rays. At the time, because it had been so cold, it was assumed that the cooling effect was greater than the warming.

Now, thanks to proper scientific investigation, we know that the warming effect is greater, which is why there is so much current activity dedicated towards cutting greenhouse-gas emissions in order to combat global warming. Yet even if we accept that warming is taking place, the precise contribution of greenhouse gases is difficult to assess. We are, after all, still coming out of the Little Ice Age that brought extremely cold spells of weather at various periods between about 1550 and 1850. Quite how much of the current climate change is just a continuation of the post-Little-Ice-Age warming-up process, and how much is due to greenhouse gases is unclear.

88. How much of the current climate change is due to human activity?

One of the vogue words of the moment is 'anthropogenic', meaning 'It's all our fault'. Yet even if we just consider the warming caused by increased greenhouse-gas levels, it is difficult to assess the human contribution with any confidence of accuracy. Burning fossil fuels certainly contributes to increasing CO_2 levels, but carbon dioxide is far from being the only, or even the worst, of the greenhouse gases. Other leading offenders include water vapour, which has its origins in a number of natural processes, and methane ($\rightarrow$ PLANTS 385), which may have the strongest effect on climate of any greenhouse gas.

Human farming practices certainly contribute to the production of methane, via the burping of cows and sheep. But we should not forget the world's 250 trillion termites, whose remorseless chomping also makes a significant contribution.

COFFEE

89. Who invented the coffee grinder?

The coffee plant is native to Ethiopia, but the first evidence of coffee beans being turned into a beverage comes from fifteenth-century Yemen. The fashion for this black, bitter drink spread across the Middle East and the Mediterranean, reaching Europe in the late sixteenth century. Although hand-operated spice mills had been in use since the 1400s, coffee beans continued to be ground using the more basic technology of mortar and pestle, or by millstones. Even as late as 1620, when the Pilgrim Fathers sailed for America on the *Mayflower*, all they brought with them for grinding coffee was an adapted mortar-and-pestle device.

In the 1660s a certain Nicholas Book, 'living at the Sign of the Frying Pan in St Tulies Street' in London, publicized himself as the only man known to make mills that could grind coffee to powder, but he was not necessarily the inventor of the machine he manufactured. The first US patent for a coffee grinder was issued in 1798 to Thomas Bruff of Maryland, who, when he was not grinding coffee, was Thomas Jefferson's dentist.

90. Does drinking coffee make us more alert?

The presence of caffeine in coffee has given the beverage an enduring reputation as a stimulant. However, research published in 2010 suggests that we may all have been mistaken. In the experiment the subjects, who included both coffee-drinkers and non-coffee-drinkers, were all asked to avoid caffeine for 16 hours. They were then given either a caffeine capsule or a placebo, and later a slightly higher dose or another placebo. Subjects then took a personality test to measure their emotional state and alertness.

The results showed that caffeine did not improve the alertness of either group, though some of the non-coffee-drinkers reported headaches and showed increased anxiety. Heavy coffee drinkers who had been given placebos, however, showed a lower level of alertness and also reported headaches.

The results seemed to show not that coffee makes us more alert, but that a lack of coffee makes coffee-drinkers less alert. A coffee-drinker's morning cup only serves to counteract the caffeine withdrawal symptoms that have built up overnight. The effects of caffeine seem to be more complex than had been thought.

91. Why does caffeine destroy the regularity of spiders' webs?

In the early 1950s the Swiss pharmacologist Peter Witt began a fascinating series of experiments on the changes in the pattern of spiders' webs when the creatures spinning them were under the influence of drugs. Different drugs produced different deformities in web pattern, and Witt developed the theory so well that he could identify which drug the spider had taken from a few simple measurements of the resulting web.

By the 1990s the idea had been taken up by others, including NASA, and various experiments showed that in general the more toxic a drug was, the greater the distortion produced in the web pattern. To the surprise of researchers, however, the most devastating effect on web pattern was produced by caffeine. The web produced by the caffeine-fed spider lost all regularity and looked like just a haphazard collection of strands.

The psychiatric literature contains many accounts of caffeine-induced psychosis in people with particular sensitivity to its effects. It may be that a similar effect serves to wreck spiders' webs.

See also SPIDERS 446

First come I; my name is Jowett.
There's no knowledge but I know it.
I am Master of the college:
What I don't know isn't knowledge.

Henry Charles Beeching (1859–1919).
('Jowett' was Benjamin Jowett, influential Master of Balliol College,
Oxford, and a translator of Plato.)

COMPOSERS

92. At what speed did J.S. Bach intend his compositions to be played?

Until the German inventor Johann Nepomuk Maelzel patented the metronome in 1815, composers could only give vague guidance on the speed at which their compositions should be performed. However, the early metronomes, although useful in ensuring regularity of tempo, were not that accurate in delivering a specified number of beats per minute. Whether such regularity is a boon or an unwanted intrusion on a performer's interpretative freedom has always been a matter of debate. The nineteenth-century German composer Johannes Brahms, for example, was quoted as saying: 'I am of the opinion that metronome marks go for nothing. As far as I know, all composers have, as I, retracted their metronome marks in later years.'

Before the advent of the metronome, composers used Italian terms such as *allegro* ('lively'), *lento* ('slow') and *andante* (literally 'going, moving', i.e. at a walking pace) to convey their intentions, while Handel frequently used the term *tempo ordinario*, suggestive of some sort of 'standard' speed. Bach never used that phrase, but some have suggested that the time signatures of his compositions, such as 3/4 or 3/8, were intended to give information on the basic tempo. Others believe that Bach was perfectly happy for performers to choose their own tempi, and his use of the usual Italian terms was an indication only of the relative speeds of one section compared with another. The American 'Bach tempo scientist' Cory Hall, for example, has concluded that *allegro* should be twice as fast as *lento*, and that *moderato* should be exactly midway between them.

To appreciate the freedom some performers feel in interpreting Bach's tempi, one has only to compare Glenn Gould's 1955 recording of the Goldberg Variations with his 1981 version: the opening aria in the later recording is played at half the speed of the earlier version. Who can tell which Bach would have preferred? Maybe he would have loved them both – or maybe he would have been quite horrified.

93. Who wrote the libretto for Haydn's *Creation*?

In 1795, near the end of Joseph Haydn's second visit to London, the violinist and impresario Johann Peter Salomon gave him the libretto in English of an oratorio on the subject of the Creation. Haydn, whose English was not good, had it translated into German by Baron Gottfried van Swieten, from which it seems to have been translated back into English. The original version, a lengthy poem entitled *The Creation of the World* based on texts from Milton's *Paradise Lost* and the Bible, is now lost, and Haydn later said that he could not remember the name of the author.

The music was written to be performed either in English or in German, and when the work was first published, both versions were given. The English version, however, came in for a great deal of criticism. For example, the poet Anna Seward (the so-called Swan of Lichfield), wrote in 1802: 'It is little wonder that the words translated from the German almost literally into English, should be neither sense nor grammar, nor that they should make wicked work with Milton.' Here is an example of the English libretto Seward so abhorred:

> *Now vanish before the holy beams*
> *the gloomy dismal shades of dark;*

the first of days appears.
Disorder yields to order the fair place.
Affrighted fled hell's spirits black in throngs;
down they sink in the deep of abyss
To endless night.

It is now impossible to tell whether the stilted language and grammar are the fault of the unknown English author of *The Creation of the World*, or whether they should be blamed on whoever it was who translated Swieten's German back into English.

94. Why did Schubert leave his Eighth Symphony unfinished?

Franz Schubert died in 1828 at the tragically early age of thirty-one, so it is hardly a surprise that he left much work incomplete, but his great Unfinished Symphony is still a puzzle. (What is surprising and hugely impressive is that, like Mozart, he wrote so much in such a short life.) Schubert began work on his Symphony No. 8 in B Minor in 1822 and wrote two movements, both fully scored for orchestra, together with the piano score of a third movement, a scherzo, of which the first two pages only were completed for orchestra – and that is all. At the time it would have been conventional for a symphony to have four movements, but there is no evidence that Schubert even started work on a finale, despite the fact that he lived for another six years, during which he completed a great deal of work, including the whole of his Ninth Symphony.

To add to the mystery, Schubert gave the incomplete score to his friend Anselm Hüttenbrenner in 1823, who did not reveal its existence until thirty-seven years after Schubert had died. Why Hüttenbrenner waited so long – and why certain pages appear to have been torn from the manuscript after the beginning of the scherzo – have given rise to a certain amount of speculation. The completed movements received their first performance in Vienna in 1865, with the last movement of the composer's Third Symphony tagged on as a finale.

Various theories have been put forward as to why Schubert failed to complete the work: he intended to complete it later; he did complete it, but the original score went missing; or he simply ran out of inspiration. A number of attempts have been made in modern times to complete the symphony, some inspired by the competition the Columbia Gramophone Company held in 1928 to mark the centenary of Schubert's death.

95. Did Schubert die of syphilis?

The official cause of Franz Schubert's early death was typhoid fever, yet there has long been a theory that the true cause was syphilis. The composer is thought to have contracted the disease from a prostitute in 1822 (though the sex of the prostitute has always been a matter for speculation), but he was supposedly cured in 1824. When Schubert's body was disinterred in 1863 (and again in 1888), to 'secure the mortal remains against further decay', it is said to have shown little sign of tertiary syphilis. However, the symptoms shown by the composer shortly before his death are consistent with mercury poisoning, and at the time mercury was often given as a treatment for syphilis.

96. Who was the woman Beethoven addressed as 'My Immortal Beloved'?

When Ludwig van Beethoven died in 1827, some letters were found among his effects, scrawled in pencil and clearly never delivered. Dated 6 July and 7 July, and believed to have been written in 1812 (though the year is not specified on the letters), they are addressed to 'My angel, my all, my very self' and 'My Immortal Beloved', and consist of an outpouring of passion.

'Oh, where I am, you are with me – I will see to it that you and I, that I can live with you.' he writes. 'No other woman can ever possess my heart – never – never... Oh God, why must one be separated from her who is so dear?' He ends, 'Oh, do continue to love me – never misjudge your lover's most faithful heart.' Nobody knows who was the intended recipient of all this passion, or why Beethoven never sent the letter. However, the wording of the letter suggests a relationship of long standing, and the 'Immortal Beloved' has been identified with a number of different women.

The favourite candidate is Josephine Brunsvik, with whom the composer had been passionately involved over a long period and to whom he had written more than a dozen other love letters. But several others also enter the frame. There was Josephine's cousin Giulietta Guicciardi, with whom Beethoven had also fallen in love and to whom he had given free piano lessons, and to whom he had also dedicated the 'Moonlight' Sonata; there was Josephine's sister Thérèse von Brunsvik; there was the Countess Marie Erdödy; there was his doctor's daughter, Therese Malfatti; there was the young singer Amalie Sebald. And there were others. For such a faithful lover, Beethoven certainly had plenty of immortal beloveds to choose from.

97. Beethoven's 'Für Elise' is one of the most popular pieces of piano music of all, but who was Elise?

The delightfully simple piano piece called 'Für Elise' ('for Elise') was composed by Beethoven around 1810, so what could be more natural than for the forty-year-old composer to have written it for the beautiful eighteen-year-old woman he had just fallen in love with and later proposed to?

The only problem with that theory is that her name wasn't Elise, but Therese. Nevertheless, Therese Malfatti, the daughter of Beethoven's doctor, has until recently been the prime suspect as the dedicatee of 'Für Elise' – and has also been suggested as Beethoven's

'Immortal Beloved' (→ 96). There is even a story that Beethoven got so drunk the night he intended to perform the piece and then ask for Therese's hand in marriage that the dedication 'Für Therese' that he scrawled on the manuscript he gave her was so illegible that it was later misread as 'Für Elise' – and the name stuck. One nineteenth-century German musicologist claimed to have seen the manuscript among the possessions Therese left at her death, but this manuscript has never been sighted again.

Recently, however, a real Elise has entered the reckoning. Elisabeth Röckel, a singer and pianist, was close to both Beethoven and his friend and colleague, the brilliant Austrian pianist and composer Johann Nepomuk Hummel. She married Hummel, but Beethoven's affection was apparently undiminished. And in 2009 the Berlin musicologist Klaus Martin Kopitz announced that his researches revealed that she was known to her circle of friends as Elise.

Others, however, have pointed out that Elise was a very common name at the time, and that Beethoven was constantly falling in love with young women.

98. Did Tchaikovsky really die of cholera?

Perhaps no other composer's death has elicited such controversy as that of Pyotr Ilyich Tchaikovsky, on 6 November 1893. The official diagnosis was cholera, but even that has given rise to conflicting theories:

(i) His death was an accident, caused by inadvertently drinking a glass of unboiled tap-water during a cholera epidemic in 1893.

(ii) He drank the tap-water deliberately in suicidal despair at his homosexuality.

(iii) The tap-water was just a convenient excuse: he actually contracted cholera from a male prostitute.

(iv) He was obliged to drink the tap-water by a 'court of honour' of his fellow alumni of the St Petersburg Imperial School of Jurisprudence, as punishment for his homosexuality.

Some have rejected the cholera theory entirely, noting that the normal quarantine regulations were not followed after his death and his body was not sealed in a zinc coffin. Instead, the composer died of arsenic poisoning, either administered by himself or by one of his doctors on the orders of Tsar Alexander III after Tchaikovsky had seduced one of his sons.

At least some of the mystery might be cleared up if Tchaikovsky's body were to be exhumed and tested for arsenic, but this is unlikely to happen. As the authoritative *New Grove Dictionary of Music and Musicians* (2001) comments, 'We do not know how Tchaikovsky died. We may never find out... '

99. What was the enigma behind Elgar's *Enigma Variations*?

There are two main theories about the enigma behind Edward Elgar's *Enigma Variations*, described by the composer as 'variations on an original theme for orchestra', and first performed in 1899. The first holds that the enigma in question is the identity of a person, while the second maintains that it is a musical theme that is implied, but never heard.

If it is a person, attention has concentrated on the thirteenth variation, which is identified only by three asterisks – in contrast to the others, which are all given sets of initials that readily identify a number of Elgar's friends and the composer himself. The main candidates put forward for Variation 13 are both ladies whose identities Elgar might have thought it delicate not to reveal. One is Helen Weaver, to whom Elgar had been engaged for eighteen months in 1883–4 before she emigrated to New Zealand; the other is Lady Mary Lygon, who sailed for Australia at about the time Elgar was writing the *Enigma Variations*. In either case, his inclusion of quotations from Felix Mendelssohn's concert overture *Calm Sea and Prosperous Voyage* in the variation might seem appropriate.

Elgar himself hinted at the other theory, stating that in addition to the theme that starts the work, on which the fourteen variations are based, 'through and over the whole set another and larger theme "goes" but is not played'. 'Auld Lang Syne', Mozart's 'Jupiter' Symphony, Bach's *Art of Fugue*, 'Rule Britannia' and 'Twinkle, Twinkle Little Star', among many others, have all been authoritatively claimed as the missing theme.

The most imaginative suggestion, however, is that the enigma is the mathematical value of π which, to three decimal places, is equal to 3.142. The evidence proposed in support of this is that the first four notes of the entire piece are notes 3, 1, 4 and 2 in the G minor scale, the key of the opening andante. In addition, Elgar marked off the first six bars of the piece with a double bar line, which supporters of the π theory claim is a deliberate ruse to draw attention to the fact that those six bars contain twenty-four black notes (all crotchets and quavers, with no 'white' minims), supposedly representing the 'four-and-twenty blackbirds baked in a pie'. He also referred to a 'dark saying' at the heart of the enigma, and what can be darker than black-note blackbirds baked in a π?

There is much pleasure to be gained from useless knowledge.

Bertrand Russell (1872–1970)

100. Would the world have come to an end if Scriabin had not cut himself shaving?

Alexander Scriabin – composer of such works as the orgasmic *Poem of Ecstasy* and inventor of the so-called 'Mystic Chord' – died on 27 April 1915 from septicaemia (blood poisoning), which developed after he cut a boil on his lip while shaving. At the time, the monomaniacal Russian – who once expressed the desire to 'possess the world as I possess a woman' – was working on a composition called *Mysterium*, a vast synaesthetic extravaganza involving the senses of sight and smell as well as hearing. Regarding this grandiose project, Scriabin proclaimed:

> *There will not be a single spectator. All will be participants.*
> *The work requires special people, special artists and a completely*
> *new culture. The cast of performers includes an orchestra, a*
> *large mixed choir, an instrument with visual effects, dancers,*
> *a procession, incense, and rhythmic textural articulation. The*
> *cathedral in which it will take place will not be of one single type*
> *of stone but will continually change with the atmosphere and*
> *motion of the Mysterium. This will be done with the aid of mists*
> *and lights, which will modify the architectural contours.*

It was to be performed in the foothills of the Himalayas in India, would last seven days and would, Scriabin believed, be followed by the end of the world and the replacement of humans by 'nobler beings'.

He was probably wrong, but we cannot be absolutely sure that we were not all saved by a shaving cut.

See also MURDER 322, MUSIC 327–8, 330

CONSCIOUSNESS

101. What is the difference between 'mind' and 'consciousness'?

The terms 'mind' and 'consciousness' have been used by philosophers, psychologists, neurologists and others for so long that distinctions between them have become hopelessly blurred. Some writers use the words almost interchangeably, some try to define them as totally distinct, others allow varying degrees of overlap. The general consensus, though, is that 'consciousness' is the state of mind that recognizes one's own existence, while 'mind' is the set of mental processes that the brain uses to process information and sensory input. Whether consciousness is entirely included in mind, however, is unclear. The American neuroscientist Antonio Damasio, in his book *Self Comes to Mind* (2010), describes consciousness as 'mind with a twist… since we cannot be conscious without having a mind to be conscious of'. Others, however, have suggested that one can lose one's mind without losing consciousness.

Until we know much more about how the brain works, it seems doubtful whether we can even define the terms sufficiently precisely to talk of the difference between mind and consciousness.

See also BRAIN 55

102. Are animals conscious?

Some would argue that language is necessary before a creature can be considered conscious. That argument seems to boil down to saying consciousness demands talking to yourself – and you can't talk to yourself without language. So if you don't believe that animals have language, they can't be conscious.

More generally, consciousness is sometimes viewed as a specifically human attribute that emerged at a particular point in our evolution and may even be taken as a characteristic of humanity itself.

A growing viewpoint sees consciousness not as an attribute that you either have or you don't, but as a factor that may be possessed in varying degrees by any creature that possesses feelings and emotions. The neuroscientist Antonio Damasio characterizes the two ends of the scale of consciousness as 'core consciousness', which is the basic sense of the here and now, and 'autobiographical consciousness', which brings in not only the present, but also a full sense of identity involving the past and the hoped-for or expected future.

Pet-lovers, of course, will say that all the above is nonsense and that Fido has feelings, and displays loyalty and love and intelligence, and that he has a mind of his own, and of course he's conscious or he wouldn't now be asking to be taken for a walk.

103. Can robots become self-aware?

The debate over artificial consciousness is similar to the arguments surrounding artificial intelligence: there is a weak version and a strong version. The weak hypothesis is that machines or robots could be programmed to behave in a manner that simulates conscious behaviour indistinguishable from that of humans. Just as computers are now being programmed to pass the 'Turing Test' of being able to respond to questions in a way that fools people into thinking they are human, there is good reason to suppose that with improving technology, computers will one day be able to pass some kind of 'Turing Consciousness Test'. But if all they are doing is carrying out programmed instructions, that, in the opinion of many, cannot count as consciousness.

The strong hypothesis is that as advances are made in data analysis, processing speeds, computational methods and decision-making techniques, true consciousness will inevitably emerge.

The debate is strongly connected to the question of whether mind and consciousness (→ 101) are a direct consequence of physical and chemical processes in the brain (→ BRAIN 55).

104. For how long does a severed head stay conscious?

Immediately after Charlotte Corday (→ FRENCH HISTORY 190) had been guillotined for the murder of the French Revolutionary leader Jean-Paul Marat in 1793, the executioner, a man named Legros, picked up her head and slapped it on the cheek. According to witnesses, an expression of 'unequivocal indignation' came over her face, which also reddened at the insult. Legros was imprisoned for three months for this breach of guillotine etiquette, but whether Mademoiselle Corday was aware of what was happening is still a matter of dispute.

Hers is one of several cases of guillotine victims supposedly showing signs of consciousness after their heads had been severed. In one instance, a gruesome experiment was carried out in which a dog's blood was pumped into the victim's severed head some time after execution, supposedly inducing twitching and movements of the lips. That and most other such cases can be explained as muscle spasms, which are purely mechanical movements rather than evidence of consciousness.

Medical evidence suggests that after sudden cardiac arrest, unconsciousness occurs within five to fifteen seconds, but brain death takes between four and six minutes. That would have allowed Corday and other guillotine victims a few seconds to have been aware of what had happened to them, but not for the face to redden in annoyance, as the blood supply would have been cut off. Some maintain, however, that the trauma of decapitation induces immediate unconsciousness. France continued to use the guillotine as a method of execution until 1977, since when controlled experimentation on this topic has proved tricky.

105. Can any part of our consciousness survive death?

For the past decade, Dr Sam Parnia of Southampton University has been studying so-called 'Near Death Experiences' (NDEs), in which patients who have suffered cardiac arrest have, after resuscitation, told of lucid visions, during which they have frequently reported information that they supposedly could not have obtained from their locations on hospital beds. In 2008 he launched the AWARE (AWAreness during REsuscitation) study, in which images were projected above hospital beds that could not be seen from the position of the patient. Accounts of NDEs from the patients are to be correlated with details of the images to look for evidence that consciousness has, in some way, left the body and explored its surroundings.

Writing in 2008, the scientist and parapsychology researcher Dr Sue Blackmore stated: 'If human consciousness can really leave the body and operate without a brain then everything we know in neuroscience has to be questioned.' She welcomed Dr Parnia's research as a scientific attempt to investigate matters usually left to philosophers and theologians.

COSMOLOGY

The development of our understanding of the way the universe works is perhaps the greatest intellectual achievement of humanity. Thanks to our understanding of the laws of physics, we can obtain precise information about stars and galaxies so far distant that even the light from them takes billions of years to reach us. Newton's laws of motion, modified by Einstein's theory of relativity, enable us to work out their mass, and analysis of the spectrum of the light we receive from them gives us information about their chemical composition.

Considering the fact that the human brain necessarily evolved with the prime purpose of improving our chances of survival in competition with other creatures on the plains of Africa, it is remarkable how flexible its thinking power has become. Even though we may have detected the galaxy now known as Abell 1835 IR1916, which is 13.2 billion light years away, there is, however, still a great deal we do not know, between here (→ EARTH 141–50) and the furthest reaches of the universe.

106. Where have all the neutrinos gone?

Ever since the British physicist J.J. Thomson discovered the electron in 1897, it has become increasingly difficult to keep pace with the proliferation of fundamental particles. In the good old days of ignorance, atoms were supposedly indivisible – indeed, the word itself comes from an ancient Greek word meaning 'that which cannot be divided'. The discovery of the electron was followed by the realization that most of the mass of the atom was in a tiny central nucleus, around which the electrons oribited. The nucleus turned out to be made up of protons (discovered in 1919) and neutrons (discovered in 1932). Yet as if splitting the atom into protons, neutrons and electrons was not enough, electrons proved to be just one of a number of different particles called leptons, while we now know that protons and neutrons are themselves made up of smaller particles, called quarks. Another class of particle, called gauge bosons, also play a role in the atom. But more of that later, for the question of neutrinos is what is bothering us at the moment.

Neutrinos are very like electrons, but without any electric charge. They whizz along at close to the speed of light, and hardly interact with anything. Every second, trillions of neutrinos pass through us and we do not even notice. Their existence was first postulated in 1930 as a means of accounting for the apparent loss of energy and momentum when an atomic nucleus decays. But it was not until 1956 that a neutrino was finally detected experimentally.

Finding a balance between theory and experiment, however, has proved even more elusive. The number of neutrinos produced by atomic reactions in the Sun ought to be predictable from measurements of the energy received, yet the number of neutrinos detected falls well short of the prediction. That disparity could be accounted for if there are different types of neutrino, some with mass and some not. But how many types are there? Measurements of energy from the Sun and other cosmic forces give us one answer, direct detection of neutrinos suggests another, while the so-called 'Standard Model' of particle physics gives a third. It has been suggested that the elusive 'dark matter' that seems to hold the universe together (→ COSMOLOGY 107) may consist of neutrinos, but until we have reconciled the question of the number of neutrino types, we remain in the dark about their role in dark matter – which brings us to the next question...

107. What is dark matter, and where does it come from?

In his hugely influential work *The Structure of Scientific Revolutions* (1962), the American physicist and philosopher Thomas Kuhn drew a distinction between science's usual plodding pace of advance, involving incremental additions to the body of knowledge, and the occasional revolutionary 'paradigm shift' that necessitates tearing down and replacing the established theories with something radically different. Copernicus's ideas about the Solar System, Newton's laws of gravity and motion, Einstein's theory of relativity and the entire field of quantum mechanics are examples of such paradigm shifts. Kuhn suggested these upheavals are precipitated as much by social, cultural and historical factors as they are by purely scientific ones.

Some moments in scientific history, however, seem particularly ripe for a paradigm shift, when the old theories are finding it more and more difficult to explain new observations, and the entire structure of current assumptions is beginning to creak at the joints. Astronomy and cosmology are now going through such a phase, though the roots of this revolution go back more than seventy-five years.

In 1934 the Swiss astronomer Fritz Zwicky discovered something inexplicable about the orbital speed of clusters of galaxies. The motion of such clusters must be determined by gravity, but there simply was not enough known mass in the clusters to account for the gravitational attraction that clearly existed. In fact, the amount of observable matter in one cluster was out by a factor of more than one hundred. To get around this anomaly, Zwicky came up with the radical idea of 'dark matter', which could not be seen and did not interact with electromagnetic radiation but had mass and therefore exerted a gravitational pull. This dark matter inhabited what had previously been thought of as empty space – which is therefore not empty at all.

Zwicky's idea was difficult to test as his dark matter was, almost by definition, undetectable. So the theory remained in the background until the 1990s, when more and more observations appeared to challenge the existing theory and demand the existence of dark matter. Not only was dark matter necessary to explain the rotation of galaxies as we currently see them, it was also needed to explain how the galaxies formed in the first place. As our picture grew of what must have happened in the first instant after the Big Bang when matter itself began to be formed, it became increasingly clear that the entire universe would not have held together without the existence of dark matter. Indeed, it seemed that dark matter must account for around 80 per cent of the matter in the universe, if not more.

In December 2009 researchers involved in the Cryogenic Dark Matter Search at Stanford University announced that they might have detected direct evidence of dark matter (though they admitted there was a 23 per cent chance it was something else). But what dark matter is and where it came from still remain a mystery – though perhaps not quite such a mystery as the next question...

108. What is dark energy and where does it come from?

In 1998 our view of the universe began to get even weirder when measurements from distant galaxies led to the conclusion that they were moving away from us faster than had previously been suspected. The conclusion was inescapable: not only is the universe expanding, but its rate of expansion is increasing. This was similar to the problem that had caused Einstein to introduce his 'cosmological constant' almost a century ago (→ EINSTEIN 164) – although his intention in so doing was to keep the universe in a steady state. Einstein's gravitational theory became tidier when it became accepted that the universe was not in a steady state, but rather expanding, making the cosmological constant unnecessary. But nothing had predicted an accelerating rate of expansion.

Einstein's theories, however, once again came to the rescue. Just as his famous $E = mc^2$ had shown an equivalence between matter and energy, it was reasoned that there must be a similar equivalence between dark matter and dark energy. The expansion rate of the universe could then be accounted for by the pull of dark energy. Quite where all this dark energy comes from is just as much of a mystery as dark matter, but the latest estimate is that the entire mass-energy of the universe consists of 72.8 per cent dark energy, 22.7 per cent dark matter and only 4.6 per cent what we think of as ordinary matter – which is therefore not ordinary at all, but rather rare, consisting of less than one-twentieth of the entire universe.

109. What is the true value of the Hubble constant?

In the 1920s the American astronomer Edwin Hubble came up with a simple law giving the relationship between the velocity V at which distant galaxies are moving away from us and their distance D from the Earth:

$$V = H_0 D$$

The term H_0 is known as the Hubble constant, and the rate at which the universe is expanding depends on its value. The age of the universe also depends on the Hubble constant, as Hubble's equation also allows us to run the expansion backwards to determine when everything in the universe was in the same place and the Big Bang (which was also Hubble's idea) happened.

The velocity of galaxies may be determined from the red shift in the light we receive from them, but we can be less sure in our measurements of their distance, which accounts for the fact that different values have been given for H_0. The recent discovery of an accelerating rate of expansion of the universe also suggests that the Hubble constant may not be constant after all, and astronomers have taken to calling it the 'Hubble parameter' instead.

Current measurements of H_0 suggest that the universe is between 13.64 and 13.86 billion years old, but that could change as our theories of variations of H_0 over time are modified.

110. Are there galaxies so far away, and moving away from us so fast, that we can never know they exist?

Putting Hubble's law and Einstein's special relativity together tells us that not only are some galaxies moving away from us at close to the speed of light, but that space itself is expanding, making their apparent speed of recession greater than that of light. Pursuing this idea leads to the concept of an 'observable universe' about 46 billion light years across. This raises the question of whether there could be galaxies ever further away, moving away from us faster than light itself, from which the light has not had time to reach us since the Big Bang itself – and never will have enough time to do so.

111. Has the speed of light always been the same?

Much of Einstein's theory of general relativity relies on the speed of light in a vacuum being a fundamental constant, unchanged by place or over time. That theory is consistent with many measurements and observations over a long period, but it has been suggested that some of the anomalies outlined above could be accounted for by variations in the speed of light over the life of the Universe. No experimental evidence has yet been found to support that idea, but some scientists are considering it to be a possibility...

112. Is the speed of light now the same everywhere?

...and even if the speed of light has always been the same in our corner of the universe, a variance in its value in distant galaxies could also account for discrepancies between theory and measurement.

113. What was the so-called 'WOW!' signal detected on 15 August 1977 by scientists searching for extraterrestrial intelligence?

On 15 August 1977 Dr Jerry R. Ehman, working on the Search for Extra-Terrestrial Intelligence (SETI) project at Ohio State University, picked up a radio signal that was quite unlike anything expected from outer space. Thirty times louder than normal background sounds, it lasted for 72 seconds. Seeing the computer printout giving details of the signal, Ehman just circled the anomalous details – which bore all the hallmarks of an artificially generated signal – and wrote the word 'WOW!' by them. The signal was never repeated, or anything similar picked up.

Some have seen this one signal as evidence of intelligent life elsewhere in the universe. Others have suggested it was simply a signal originating on Earth that had been reflected back by a piece of space debris. But the WOW! signal has never been explained.

114. What is the cause of gamma ray bursts, a phenomenon that could be devastating for our planet?

From billions of light years away, flashes of gamma rays, probably associated with massive explosions in distant galaxies, have reached Earth, lasting anything between a few milliseconds and several minutes. A burst is followed by an afterglow, and is estimated to release as much energy as the Sun will give out in its entire lifetime. Because of their huge power, gamma ray bursts (GRBs) are thought to be associated with such cataclysmic galactic events as supernovae, which occur when a high-mass star collapses to form a neutron star or a black hole.

All GRBs so far detected have originated outside our own galaxy, the Milky Way – which may be fortunate as it has been suggested that any GRB in our galaxy and beamed in our direction would release enough electromagnetic energy to extinguish all life on Earth. No GRB has yet been connected to a specific supernova event that might have caused it, and no one has yet explained the mechanism by which GRBs are generated.

See also BLACK HOLES 50–3, EINSTEIN 164, THE MOON 311–12, PHILOSOPHY 367, PLANETS 374–81, THE SOLAR SYSTEM 436–9, THE SUN 454–5, THE UNIVERSE 458–66

Science is the belief in the ignorance of experts.

Richard Feynman (1918–88)

DINOSAURS

The problem with dinosaurs is that they lived a very long time ago – between 250 million and 65 million years ago – and the vast majority of our knowledge of them is based on fossilized bones and skeletons, which are all that have survived the ravages of time.

115. Were the dinosaurs cold-blooded, like all today's reptiles?

Until the 1970s the view that dinosaurs, like other reptiles, were cold-blooded was almost unquestioned. Cold-bloodedness (something of a misnomer) means that an animal has a variable body temperature, which it has to control by external means (such as basking in the sun to warm up). In contrast, warm-blooded animals, such as birds and mammals, maintain a constant body temperature using internal thermoregulation mechanisms.

Since the 1970s, some palaeontologists have pointed to a number of factors that suggest dinosaurs may have been warm-blooded after all. Unlike modern reptiles, which either crawl or walk with their limbs extended out to the side, one of the main groups of dinosaurs walked with their limbs directly beneath their bodies, like birds and mammals. Dinosaurs also had large ribcages that could have held mammal-like hearts and lungs, and their bones had channels for rapid blood circulation similar to those found in the bones of warm-blooded animals. Supporters of cold-blooded dinosaurs, however, argue that in the largest dinosaurs a warm-blooded metabolism would be likely to cause internal overheating and immediate death.

A recent study, however, has suggested that the bigger a dinosaur was, the warmer its blood, suggesting that the question may have no single answer.

116. Why were some dinosaurs so large?

Dinosaurs were, on average, much larger than any land animals seen today. Several theories have been put forward to explain this:

(i) During the era of the dinosaurs, the Earth was covered by such lush vegetation that even the largest plant-eaters had no problem finding enough food. This being the case, the bigger the animal the greater the evolutionary advantage, as large size is a good defence against being killed and eaten by one's carnivorous colleagues. In a sort of arms race, the carnivores grew big for the same reason: the bigger they were, the bigger the prey they could take on.

The suggestion that large vegetarian dinosaurs were better equipped to reach the higher branches of plants is less convincing, as there is no reason to believe there was a shortage of food at the time.

(ii) Those who believe that dinosaurs were cold-blooded maintain that is why they were so big: cold-blooded creatures with greater mass are better able to maintain constant body temperature, warming up during the day and cooling slowly at night. In smaller cold-blooded creatures, `the larger ratio of surface area to mass may have made this more difficult in the prevailing climatic conditions.

(iii) Those who believe that dinosaurs were warm-blooded also claim that is a factor in their large size. They point out that the uninsulated skin of dinosaurs was a source of heat loss, and the lower ratio of surface area to mass in larger dinosaurs would have protected them from the chilling effects that would have been dangerous to a warm-blooded creature.

To conclude: large size may have kept cold-blooded creatures cool, or it may have kept warm-blooded creatures warm...

117. What was the lifespan of dinosaurs?

To answer this question, we need to know two things about the remains of a dinosaur. Firstly, how old was it when it died? And secondly, did it die of old age? Although we can sometimes tell from evidence of injury that a dinosaur met a violent end, it is generally impossible to answer either of these questions.

Estimates of the lifespan of dinosaurs are therefore based on what we know, from the evidence of present-day animals, of the connection between size and lifespan and the possible role of metabolism. Calculations based on these factors have resulted in estimates from a respectable 75 years to an awesome 300 years for the lifespans of dinosaurs – a very broad range indeed.

118. What colours were dinosaurs?

Until recently, we knew nothing of the skin of dinosaurs. Depiction of their colour was left to the imagination of illustrators, who tended to go for a mixture of the green of lizards and the grey or murky brown of elephants and rhinos. In 2009, however, remnants of skin and 'proto-feathers' were found on dinosaur remains in China. These, and certain other features of dinosaur remains, have encouraged a belief that both skin and 'proto-feathers' were part of the creature's display mechanism to improve its breeding prospects.

Whether dinosaurs of the opposite sex liked bright colours is a matter for pure speculation, but at least the find has encouraged illustrators to open their palettes to a more striking and dazzling range in recent years.

The fundament upon which all our knowledge and learning rests is inexplicable.

Arthur Schopenhauer (1788–1860)

DISEASE

119. What was sweating sickness?

Sweating sickness, also known as the 'English sweate', was a highly virulent disease that struck England in a series of epidemics between 1485 and 1551, after which it apparently vanished, with the last recorded case in 1578. The onset of symptoms was sudden and dramatic, starting with anxiety, followed by violent cold shivers, giddiness, headache and excruciating pains in the neck, shoulders, arms and legs, with severe fever and great exhaustion. Death often occurred within hours. Its virulence was described by Dr John Caius in *A Boke or Counseill Against the Disease Commonly Called the Sweate, or Sweatyng Sicknesse* (1556). In this, Caius reports that the disease

> *immediately killed some in opening their windows, some in*
> *playing with children in their street doors, some in one hour,*
> *many in two it destroyed, & at the longest, to they that merrily*
> *dined, it gave a sorrowful supper. As it found them so it took*
> *them, some in sleep some in wake, some in mirth some in care,*
> *some fasting & some full, some busy and some idle, and in one*
> *house sometime three sometime five, sometime seven sometime*
> *eight, sometime more some time all, of the which, if the half in*
> *every town escaped, it was thought great favour.*

At this distance of time, there is little prospect of scientists ever establishing the causative organism, although it was clearly some kind of infection.

120. What is the cause of autism?

The condition (or set of conditions) known as autism was only named in 1943, and is still a long way from being understood. Characterized mainly by impairments in social interaction and communication, autism is seen by some as a distinct mental disorder and by others as an extreme point on a spectrum that begins with the mildly obsessive condition known as Asperger's syndrome. Some suggest that autism should not be treated as a disorder at all, but merely as an extreme but acceptable point in the range of normality.

Studies have shown that autism has a strong genetic component, with several genes identified as playing a possible part, though no single gene appears to be responsible for autism on its own. An imbalance in the brain's production of certain hormones has also been associated with the condition, while some blame post-natal factors, such as diet, medication or some other external factor.

It seems increasingly clear that autism is due to a complex set of distinct causes that may be connected – but which are not necessarily so. There is still much unravelling to be done.

Must helpless man in ignorance sedate
Roll darkling down the torrent of his fate?

Samuel Johnson (1709–84)

121. Why are engineers more likely to have autistic children and grandchildren than non-engineers?

To test the theory that autism has a genetic component, in 1997 the British psychologist Simon Baron-Cohen and others compared the family histories of children with autism. Their findings showed that in a sample of 919 families with a child with autism, 28.4 per cent had either a father or grandfather who was an engineer, compared with only 15 per cent of families without autism. Later studies also showed that students of mathematics and physics were more likely to have autism in their families than students of subjects demanding a high degree of social understanding. Perhaps social interaction is simply too fuzzy or too imprecise a skill to catch the interest of the methodical autistic brain.

122. Is there a connection between cholesterol intake and arteriosclerosis?

Cholesterol is an essential component of cell membranes in mammals, and also plays a part in various metabolic processes. But eating a diet rich in such things as cheese, eggs and fatty meat can lead to high levels of cholesterol in the blood. There is no doubt that arteriosclerosis (a build-up of calcium on the insides of artery walls) and atherosclerosis (a similar build-up of fats) are linked to high levels of cholesterol in the blood and lead to heart attacks. But whether the diseases are caused by high cholesterol intake is still open to question. An alternative view is that the diseases themselves interfere with the body's mechanism for breaking down cholesterol, and thus it is the diseases that are responsible for the cholesterol build-up, and not the other way round.

There is a (no doubt apocryphal) story of an American who heard that the incidence of heart disease in the USA is far greater than in Japan, where the diet is low in cholesterol. But there is also a low level of heart disease in France, where cholesterol intake is as high as anywhere (think of those 246 varieties of cheese). The American concluded that the cause of heart disease was not cholesterol but speaking English. He was probably wrong about English, but he could have been right about cholesterol.

123. What causes eating disorders?

Behavioural, biological, biochemical, emotional, psychological, interpersonal and social factors have all been blamed for eating disorders such as bulimia and anorexia nervosa, but even cases with apparent physical causes still pose many questions. In some individuals, a relationship has been shown between eating disorders and chemicals in the brain that control hunger, appetite and digestion, but whether it is the eating disorder that causes the chemical imbalance or vice versa is unclear. Eating disorders have also been frequently found to run in families, raising questions of a possible genetic component.

124. What causes pimples and acne?

Pimples occur when a pore or hair follicle collapses in on itself and blocks the secretion from the sebaceous gland of sebum, an oily secretion that lubricates the hair and skin. The trapped sebum builds up into an oily spot. Various factors have been identified as causes of pimple formation, such as poor diet (including vitamin deficiency), hormonal imbalance (which is why teenagers are particularly prone to spots) and stress. What is not clear is why one sebaceous gland should give rise to a pimple while another does not.

125. Why are some babies born prematurely?

Around 7 per cent of babies in the UK are born prematurely (before the thirty-seventh week of pregnancy). Estimates in the USA vary between about 8 and 12 per cent. Various causes have been identified, including multiple pregnancy, pre-eclampsia (high blood pressure usually associated with problems in the placenta), haemorrhage, illness in the mother, foetal abnormality and cervical incompetence (premature widening of the cervix). In around 40 per cent of cases, however, the cause is unknown. But given that the immediate cause of a woman going into labour at full term is also unknown, this is perhaps not surprising.

126. How did AIDS and HIV originate?

The first reference to acquired immune deficiency syndrome, abbreviated to AIDS, in medical literature was in 1982. The disease had first been identified among gay men, but was later also found in other groups, such as haemophiliacs (who depend on frequent blood transfusions) and intravenous drug users. Cases of the disease itself are thought to date back to the mid-1970s, and by the mid-1980s it had spread to every continent except Antarctica. The discovery in 1983 of the human immunodeficiency virus (HIV), which is now generally accepted as the cause of AIDS, gave direction to a search for the origins of the disease.

In 1999 a type of simian immunodeficiency virus (SIV) that was very similar to HIV was found in a frozen sample of blood taken from a captive chimpanzee that had originated in West Africa. Tracing back the origins of that SIV virus led to the conclusion that it had come from two other strains of SIV and had at some point crossed species from chimps to humans. How it did so has been the subject of diverse theories:

(i) It may have spread to humans through the killing and eating of chimpanzees.

(ii) It may have entered through cuts into the blood stream of hunters who had killed the chimpanzees.

(iii) It may have been spread through the testing of an oral polio vaccine that had been cultivated in cells from the kidneys of chimpanzees.

Conspiracy theorists (with no tangible evidence) have also suggested that AIDS may have been part of a secret CIA project, either deliberately designed to wipe out gay men, or as a general biological weapon.

127. Where does the ebola virus live between human outbreaks?

Ebola is one of the most contagious and virulent diseases known to medical science. It can be transmitted by no more than a handshake, and once infected the victim suffers massive internal bleeding, as the walls of the blood vessels are destroyed. There is no known cure, and between 50 and 90 per cent of those infected die.

The first identified outbreak was in 1976 in Sudan and in a nearby region of Zaire. The disease is named after the Ebola River Valley in the Democratic Republic of the Congo (formerly Zaire). Others followed, but nobody knew where the virus lurked between human outbreaks. Bats have long been suspected, as they have frequently been found at the site of ebola outbreaks. Insects, birds and non-human primates, which are all known to have died from the disease, have also fallen under suspicion.

As with AIDS, it has been suggested that human epidemics have begun with the eating of diseased bush meat (as the flesh of hunted wild animals is known in Africa). However, since ebola is as quickly fatal to other primates as it is to humans, it seems unlikely that they are a reservoir for the virus between outbreaks. The search for its hiding place continues.

128. Why has the incidence of asthma in humans increased so much in recent decades?

World Health Organization studies report significant increases in the incidence of asthma in a number of different countries over recent decades. The rate among US children increased from 3.6 per cent in

1980 to 9 per cent in 2001, while the rate in Switzerland increased from 2 per cent around thirty years ago to 10 per cent today. This may partly be due to an improvement in diagnosis rates. From 1930 until 1950, asthma was officially considered a psychosomatic illness and treated as a psychological rather than a medical ailment. On the other hand, asthma has long been associated with allergic conditions, which may be exacerbated by inner-city living, which is on the increase. Other environmental and genetic factors are also known to be involved, but the reasons for the worldwide increase in the disease are not fully understood.

129. Does cold weather increase your chance of catching a cold?

This is a question on which we seem to be constantly changing our minds. Before we understood about germs, the accepted folk wisdom was that cold weather brought colds. That view was then dismissed as an old wives' tale, and the increased prevalence of colds in winter was explained by the fact that people spend more time indoors during cold weather, thus increasing their chance of catching an infectious disease from somebody else.

In 2005, however, an interesting experiment was performed at the Common Cold Research Centre in Cardiff in which ninety people immersed their feet in a bowl of ice-filled water for twenty minutes, while a control group kept their feet in an empty bowl for the same period. Over the next five days, 29 per cent of the group who had had their feet in icy water developed cold symptoms, compared to only 9 per cent of the control group.

The explanation seems to be that coldness may not cause the common cold, but it may decrease the body's resistance by reducing the supply of white blood cells – a primary defence against the various viruses that cause the common cold. But since the specific viruses responsible for the common cold have not been identified, we cannot be sure of that.

130. Can you catch a cold by kissing?

It used to be thought that kissing was an effective way of spreading colds, but an experiment conducted at the University of Wisconsin in 1984, in which people without colds kissed people with colds, resulted in only one in thirteen of the healthy kissers catching the disease. The official line since then has been that cold viruses are not carried on the lips or in the saliva, but live in the mucus in the nose. So kissing is fine, but Eskimo-style nose-rubbing is out.

On the other hand, close contact with a sufferer has always been held to carry the risk of catching a cold, as the virus may easily be transferred from nose to hand or into the air. So kissing is only all right if you can do it without being in close contact with the person kissed – particularly, one might think, close contact with any part of the anatomy near their nose.

131. Will disease ever be eradicated?

We have eradicated smallpox. We may be on the verge of eradicating polio. We are on the point of eradicating rinderpest in cattle. We thought we were well on the way to eradicating tuberculosis, until the emergence of drug-resistant strains. So will we ever eradicate all diseases? There seems to be no reason in principle why we should not, one day, develop a way of eliminating all the things that interfere with the functioning of the human body, unless, of course, disease is in some way essential to the continuing existence of the human race. After all, if being disease-free is an evolutionary advantage, which we might naively assume to be the case, then why have we not evolved that way? The disease-creating organisms always seem to be one step ahead of us in the evolution game.

See also ANTHROPOLOGY 26, ARMADILLOS, 29, BIOLOGY 43, COMPOSERS 95, 98, FOOTBALL (AMERICAN) 187, GARLIC 205, MEDICINE 282–7, THE MIDDLE AGES 303, SMOKING 434–5

DNA

The discovery by Francis Crick and James D. Watson in 1953 of the double-helix structure of the DNA (deoxyribonucleic acid) molecule has proved to be one of the most far-reaching in the history of science. Darwin's theory of evolution and the science of genetics had already been with us for many years, but the discovery of the double helix showed how genetic information is encoded at the molecular level. The DNA of each individual organism provides a unique blueprint for that organism, as well as the information that makes it the species it is. Every organism has a copy of its unique DNA in every one of its cells. As the genotypes of more species and more individuals are unravelled, we are able to answer more and more questions regarding the origins and relationships of different organisms. But the explosion of information this has provided has created a vast cloud of unanswered questions, of which the following are perhaps only the most basic.

132. Is all life DNA-based?

Every independent life form so far encountered in the universe is DNA-based. Even viruses, which can only survive by invading a host cell, are based either on DNA or the closely related RNA. DNA (deoxyribonucleic acid) contains the genetic instructions for the development and functioning of all organisms, from the humblest plant or microbe to the human being. If there is another way for life to function, we haven't found it yet. Or if we have found it, we haven't recognized it, perhaps because it is so different from us. If DNA is essential for life of any form, then the question we have to ask is: What is it about DNA that makes it so special?

133. Why do some highly evolved complex organisms have less DNA in their genome than simpler organisms?

All the information needed to make a human being, or any other living thing, is contained in his, her or its genome, which comprises the entirety of the organism's hereditary information and which is encoded in its DNA. Long stretches of the genome form the heritable units known as genes, and our genes are transmitted to us on our chromosomes (organized structures of DNA and protein found in each cell). So one might think that the human being – the most complex organism we know – ought to have a longer genome, and more genes and chromosomes, than anything else.

Far from it. The human genome consists of about 3 billion pairs of bases consisting of the structural units known as nucleotides. Just to read out the human genome would take about nine or ten years without stopping. Pretty complex, you might think, but the microscopic *Amoeba dubia* has a longer genome than us, as do the lungfish and the Easter lily.

So what about our genes? Well, we have about 25,000 genes, which is about the same number as mustard grass and not that many more than a roundworm. Finally, how do we score on chromosomes? We have a respectable 46 chromosomes (23 pairs), but carp have 104 – and even some species of potato have 48.

It may be that the organisms mentioned above need all their genetic material, but much of it may simply represent redundant leftovers from a long evolutionary process.

134. Why does the vast majority of information in our DNA appear to have no function?

As far as we can tell, the main role of DNA is to provide the information needed to construct the large variety of proteins required to build our bodies and keep them functioning. Yet it has been estimated that as little as 2 per cent of our DNA has that function. The rest used to be known as 'junk DNA', with the implication that it was like the lines of code in a lazy computer program that had been overwritten by later instructions but never removed.

More recently, attention has focused on segments of this apparently 'junk' DNA that seem to have the function of determining when genes are turned on and off. It may turn out that much of our genome has such a controlling function, or it may turn out that much of it really is junk – a no longer wanted genetic inheritance from a distant past.

See also EGYPTOLOGY 162, EVOLUTION 182–6, GENETICS 206–14, MOZART 315, PALAEONTOLOGY 355–6, PROTEINS 393, 395

DODOS

135. Is the dodo extinct?

The dodo is generally believed to have become extinct soon after 1662, when the last known specimen of this large flightless bird was sighted on the Indian Ocean island of Mauritius. Natural history, however, is littered with examples of species that have, as it were, come back from the dead. In 1938, for example, fishermen off the coast of South Africa netted a large, primitive-looking fish. This turned out to be a coelacanth, a species that had supposedly died out 65 million years ago. So what is the chance that a dodo is still around somewhere?

In 2010 the journal *Nature*, under the title 'Should we be trying to save the dodo?', reported a new statistical technique to estimate the chance of saving threatened species, or finding living examples of species thought to be extinct. The technique is based on a formula into which are entered the dates of the last ten confirmed sightings of the animal, and from this it calculates the chance that a live specimen is still lurking undiscovered somewhere. When the dodo's details are entered, it comes up with the answer that the probability that the dodo lives on is 3.07×10^{-6} – or about three chances in a million.

DOGS

136. Why are dogs even more likely to catch yawning from humans than other humans are?

A fascinating experiment was conducted at London University in 2008 to determine whether human yawning is contagious to dogs. Of the twenty-nine dogs in the study, twenty-one yawned after watching the researcher yawn. That represents 72 per cent of the sample catching the yawn, compared with only 45 per cent (or in some studies 60 per cent) reported in humans, and 33 per cent in chimpanzees. Furthermore, when the humans just opened their mouths, none of the dogs yawned, suggesting that only true yawns are contagious.

137. Are dogs psychic?

In 1983 the academic parapsychologist Helmut Schmidt performed an experiment in which a dog was rewarded with chocolates whenever a human subject made a correct guess in a test of extrasensory perception. The dog, which was described as a 'supposedly non-psychic' miniature dachshund, was thus motivated to try to influence the man's performance. The man's performance did indeed improve during the course of the experiment, but the dog was found to do even better when rewarded for making its own correct choices in an ESP experiment without having a human as an intermediary. It was concluded that further research was needed.

In 1999 Dr Rupert Sheldrake published a book entitled *Dogs That Know When Their Owners Are Coming Home and Other Unexplained Powers of Animals*, in which he presented case studies of dogs that apparently showed signs of knowing, at a distance, the movements of their owners. A number of academics have indicated that they are unconvinced.

138. Do dogs distinguish rational from irrational acts?

This question formed the title of a paper in 2010, which reported two experiments to determine whether dogs were more likely to follow directions given to them in a rational manner than those delivered in an irrational way. In the first experiment, humans pointed with their legs in the direction in which food could be found. Some of the humans had their hands occupied, which made pointing with their legs a rational act, others had free hands, which made pointing with the legs irrational (or at least unnatural). The second experiment involved trained dogs demonstrating a task that involved operating a lever. The natural doggy way to pull the lever would be to use the mouth, but the trained dogs did it with a paw, either when their mouths were holding a ball, making the use of the paw rational, or when their mouths were empty, making the paw-use supposedly irrational. The experimenters concluded that 'our results suggest that dogs do not distinguish rational from irrational acts', which contradicted the conclusions of earlier experimenters. So the question is still open.

See also THE BRAIN 58, CONSCIOUSNESS 102

DRUIDS

139. What did the ancient Druids believe?

Between 200 BC and AD 200, various Greek and Roman writers described the Druids, who to them appeared as a mysterious priestly class with great influence in Britain, Ireland and Gaul. Julius Caesar wrote that they acted as judges in Gaul, and were highly respected as intermediaries between men and gods, but all he mentions regarding their doctrine is that they believed in reincarnation – the transmigration of souls. The Druids themselves left no written records. What we do know about the Druids is that they had nothing to do with Stonehenge – a much earlier structure (→ ANCIENT HISTORY 13). The connection between the two was not made until the 17th century – and the 'Ancient Druid Order', far from being ancient, was not founded until 1717.

140. Did the Druids use a 'wicker man' to burn human sacrifices alive?

Julius Caesar also says that the Druids indulged in human sacrifice, the victims usually being criminals but sometimes innocent people. However, history is littered with examples of such allegations, intended to give the impression that the people one wishes to conquer are utterly barbarous, and in need of a civilizing hand. In Caesar's account, the victims were burned alive in a large wooden effigy, known in recent times as the 'wicker man' (as popularized in the cult 1970s film of that name). In another account, also from the first century BC, the Greek historian Diodorus Siculus, reports that the Druids made their human sacrifices by plunging a dagger into the victim's chest and forecasting the future from the way his legs twitched.

There is no archaeological evidence of wicker men, and evidence of human sacrifice is strongly disputed.

THE EARTH

141. What causes ice ages on Earth?

Over the last 2 or 3 billion years, geological evidence tells us that there have been at least five ice ages on Earth, but we do not really understand the mechanics of what causes them. One factor must be the amount of heat we receive from the Sun, which varies according to slight changes in the Earth's orbit and the angle between the Earth's axis of rotation and the plane of its solar orbit. Other factors may also affect the energy emitted by the Sun, but we do not know what they are. Another factor is the gravitational pull of other planets on the Earth's orbit, and there has also been a suggestion that the temperature on Earth may be affected by changing temperatures of the space through which its orbit takes it.

Around 1940 the Serbian geophysicist and engineer Milutin Milanković studied changes in the Earth's orbit and showed that it underwent various potentially climate-changing wobbles with periods of 21,000 years, 26,000 years and 41,000 years, now known as the Milanković cycles, which must have something to with the

Earth's periodic ice ages, though they fail to fit perfectly with the geological data.

There is also the question of continental drift and plate tectonics, which by varying the relative positions and areas of sea and land influence how much solar heat the planet retains at its surface, because sea is slower to warm and slower to cool than land. Greenhouse gases may also have played a role. It has been suggested, for example, that gases emitted by widespread volcanic eruptions 630 million years ago may have been largely responsible for the end of one ice age.

To complicate matters further, there are also glacial periods within each ice age, during which the ice sheets advance and recede. For the past 11,000 years or so, we have been in an interglacial period, with the ice sheets generally receding, but from around 1350, temperatures began to fall until the start of the seventeenth century, bringing a 'little ice age' that lasted until the early nineteenth century, a chilly period which nobody can completely explain.

142. Why did the Earth's temperature rise dramatically about 55 million years ago?

At the very end of the Palaeocene era, 55 million years ago, there was a climatic anomaly known as the Palaeocene–Eocene thermal maximum. Ocean surface temperatures worldwide shot up by between five and eight degrees Celsius and stayed that way for a few thousand years. In the Arctic, it heated up even more, to an estimated 23°C. The main effects of such warming would have been an expansion of the water in the oceans, changing the planet's ratio of sea to land. There would also have been a marked effect on the species of marine life. Geological evidence and the fossil record both confirm that such changes happened, as do chemical analyses of molecules found in the fossil shells of primitive creatures. But nobody knows what caused the huge temperature change.

143. What caused the so-called 'Tunguska event' above Russia in 1908?

On 30 June 1908 a massive explosion – about a thousand times more powerful than the atom bomb that destroyed Hiroshima – occurred in the sky near the Tunguska River in a remote part of Siberia. Hundreds of academic papers have been written on the subject, yet we are still not sure what it was. Some suggest a comet or asteroid that vaporized in the Earth's atmosphere, but it left no evidence or even a crater. Eye-witness accounts, however, told of a column of blue light almost as bright as the Sun moving across the sky, followed by a sound like heavy artillery fire. According to one account it was as if 'the sky split in two'. Seismic stations across Eurasia registered a huge explosion, while the atmospheric shock wave was detected as far away as the UK. Later investigators established that over 2,000 square km (800 square miles) of forest were flattened, with some 80 million trees knocked over by the blast. There were also numerous holes in the ground that were at first thought to be meteor craters, but no traces of meteors, or anything else from whatever exploded, have ever been found.

144. Why did the Earth's thermosphere collapse so badly in 2008–9?

Way above the Earth's surface, at a height of between 90 and 500 km (55 and 300 miles), lies the thermosphere, the highest level of the Earth's atmosphere. The thermosphere traps ultraviolet radiation from the Sun and its temperature increases the higher you get, unlike the other three atmospheric levels (the troposphere, stratosphere and mesosphere), which get colder with height.

In 2008–9, however, the thermosphere showed a dramatic shrinkage, which has not been explained. It is known that an increase in carbon dioxide in the atmosphere can reduce the size of the thermosphere; it is known that a decrease in solar activity can also have that effect; but calculations taking both of those into account still leave at least 60 per cent of the shrinkage unexplained. There is clearly some aspect of the Sun's influence on Earth that we do not understand.

145. What is the cause of the so-called flux transfer events that open a magnetic portal between the Earth and Sun, and why do these portals open every eight minutes?

Enveloping the Earth, some 70,000 km (43,000 miles) above the planet's surface, is a region of space called the magnetosphere. Only discovered in 1958 by the *Explorer 1* space probe, the magnetosphere (which isn't a sphere at all, but more of a teardrop shape) is formed by the action of the Earth's magnetic field on the free ions from the ionosphere (a part of the uppermost layer of the atmosphere) and electrons from the solar wind. Its magnetic forces, in fact, act as a protective shield against the potential ravages of the solar wind. Every eight minutes, however, a portal opens in the magnetosphere that allows high-energy particles from the Sun to flow through to the Earth. This is known as a flux transfer event, but no one yet knows what causes them or why they happen every eight minutes.

146. What was the history of the Earth before the break-up of the super-continent Pangaea?

When in 1912 the German meteorologist Alfred Wegener proposed that the Earth had once consisted of a single super-continent, the idea was greeted with great scepticism. The hypothetical continent was not even given a name until 1926, when a conference on Wegener's theory of continental drift came up with the name of Pangaea (from the Greek for 'all land') and dubbed the ocean that surrounded it Panthalassa (meaning 'all sea'). Suspicion of Wegener's theory diminished as the fossil evidence accumulated, and was entirely vindicated in the 1960s, when plate tectonics provided a plausible mechanism not only for continental drift, but also for seismic activity and volcanism. But what the Earth was like before Pangaea remains a matter for speculation.

One theory, supported by some geological evidence, is that the planet goes through a cycle in which, over a period of between 300 and 500 million years, super-continents form and then break up. But what the map of the Earth looked like during these earlier cycles, if they indeed occurred, remains largely a matter of guesswork.

147. What is the Earth's core made of?

Considering how much we now know about the rest of the universe, we know surprisingly little about what goes on deep inside our own planet. We can calculate, from our knowledge of the strength of the Earth's gravitational pull, what its mass must be; we can measure directly the density of the material near the surface of the Earth; and we can deduce from those figures that the inside of the Earth must be much denser than the part we know.

Apart from that, our main evidence of the nature of the Earth's core has come from the analysis of seismic waves caused by earthquakes. Unexpected patterns in the readings of seismographs at different places on the Earth's surface indicated that the waves were meeting some obstruction in their passage, and that is what led to the current picture of a solid inner core with a radius of about 1,220 km (760 miles), surrounded by a liquid core extending to around 3,400 km (2110 miles) from the Earth's centre.

Considerations of density and examination of meteorites thought to have formed in the same way as our planet have led to the conclusion that the inner core is composed mainly of iron and nickel, but it seems generally agreed there are also other elements there, at present unidentified. In 2006, after examining the composition of meteorites believed to be similar to the 'planetesimals' that crashed together to form the Earth billions of years ago, the Australian geologist Bernard Wood concluded that 99 per cent of the Earth's gold is missing, and that it must have sunk to the core, which must also contain most of the Earth's platinum. All we have to do to find out is to drill down about 5,000 km (3,000 miles).

148. What is the temperature at the centre of the Earth?

In Dante's *Inferno*, at the centre of the Earth lies Cocytus, a frozen lake that forms the ninth and lowest circle of Hell where traitors such as Judas are punished for all eternity. Satan himself is trapped at its centre, his constant tears freezing as they add to the icy lake.

This picture of the centre of the Earth is not supported by scientific opinion, but there is still considerable discrepancy in estimates of the temperature at the centre of the Earth, which vary between about 4,000 and 7,000°C. These estimates are based on calculating the melting point of iron at the boundary of the inner and outer core, at the huge pressure known to exist at the Earth's centre.

149. When will the next mass extinction take place on Earth?

Fossil records indicate that there have been five major extinctions on Earth in the last 500 million years, each resulting in the extinction of more than half the life forms on the planet. The most recent was about 65 million years ago and resulted in the extinction of the dinosaurs.

There is a good deal of evidence that that extinction was caused by the impact of an asteroid, though alternative explanations include climate change caused by volcanic eruptions, the evolution of mammals that ate dinosaur eggs, and a change in plant-life resulting in the herbivorous dinosaurs becoming fatally constipated. The cause of the other mass extinctions is unknown, which makes it difficult to estimate when the next one will be. Potential causes of future extinctions include human-generated environmental change, energy bursts from outside our galaxy, climate change that is outside our control, and stray asteroids crashing into the Earth. In any case, the Sun, as it burns out in about another 5 billion years, will explode into a red giant and engulf the Earth, though at the current rate of one mass extinction every 100 million years, we will have had about fifty mass extinctions before then anyway.

150. Did life on Earth come from outer space?

The fossil record indicates that life on Earth began around 3.5 billion years ago, a relatively short time after the planet was formed 4.6 billion years ago. Doubts have been cast on whether this is long enough for the complex molecules that are necessary for life to have formed from carbon, oxygen, hydrogen and nitrogen. At around the same time as life on Earth first appeared, our planet was subjected to the 'Late Heavy Bombardment', when the Earth and Moon were battered by large numbers of rocks from space. Since the universe had been around for some 9 billion years before the Earth was formed, there would have been plenty of time for the so-called 'prebiotic molecules', or even basic life forms, to have appeared elsewhere, and these could have been dumped on Earth by meteorite strikes. This theory is called exogenesis (meaning 'birth from outside'), but would need good evidence of life elsewhere in the universe to support it.

See also ANCIENT HISTORY 17, CLIMATE 87–8, EVOLUTION 182, THE PLANETS 375, THE SOLAR SYSTEM 439, WATER 470

The learning and knowledge that we have, is, at the most, but little compared with that of which we are ignorant.

Plato (424/3 BC–348/7 BC)

EARTHQUAKES

151. What is the nature of the forces that drive the movement of the Earth's tectonic plates?

Until the early twentieth century, it was generally believed that the world as we know it was quite literally set in stone. The planet might once have been a mass of gases and molten rock, but everything had now cooled down, leaving us with the unchanging land and sea as we now see it. In 1912, however, the German meteorologist Alfred Wegener came up with the theory of 'continental drift'. He proposed that the Earth had once consisted of the oceans and a super-landmass that he called Pangaea, which had split up to form the continents as we now know them (→ THE EARTH 146). Wegener also maintained that this process continues to this day – the continents are still on the move. Wegener's theory provided an explanation as to why similar or identical fossils and rock types are found in places far apart from each other – but he had no explanation of what was making the continents move.

In the 1960s the theory of continental drift was supported by a new and wider-reaching theory: that the lithosphere, the outer rocky shell of the planet comprising the crust and the upper mantle, is made up of a number of vast and mobile 'tectonic' plates. There are roughly seven major plates and seven minor plates, depending how you count them, and the lithosphere they make up sits on top of the asthenosphere, a hotter and therefore more plastic and movable layer of the Earth's mantle. Where plates are moving away from each other, new lithosphere is formed by molten rock rising from the asthenosphere below. This process is called sea-floor spreading, and accounts for the mid-ocean ridges found in the Atlantic, Pacific and Indian Oceans. At other boundaries, one plate is sliding under another, and part of the lithosphere, as it is pushed deeper, heats up and melts. There are also boundaries where the plates simply slide (or grind) past each other, with lithosphere being neither created nor destroyed.

Studies of where earthquakes and volcanic activity occur, together with aerial mapping of the sea floor, confirmed the theory, and demonstrated that it is the movement of the plates that causes these geological disturbances. Yet there is still argument about what is driving the movement of the plates.

One theory is that heat from the Earth's core causes convection currents in the asthenosphere, which lead to movement in the tectonic plates. Recent imaging of the internal structure of the Earth, however, has failed to identify the convection cells that would support this theory. Instead, plumes or channels of heat have been suggested, but they too have not been confirmed.

Another theory is that tectonic movement is at least in part gravity-driven as molten rock beneath the ocean cools and solidifies at the edge of plates, becoming more dense and slipping down beneath the neighbouring plate, causing further movement.

A third theory claims that the Earth's rotation and tides exert a force on the Earth's crust, which brings about movement of the tectonic plates.

Perhaps the answer is a combination of all of these, to one degree or another.

152. How do animals predict earthquakes?

Accounts of animals apparently predicting earthquakes have been known since ancient times. In 373 BC Greek historians recorded that rats, snakes and weasels deserted the city of Helice in large numbers just days before a devastating quake took place. Fish moving violently, chickens that stop laying eggs and bees leaving their hives have also been reported, while elephants in Sri Lanka and Thailand in 2004 were reported to have saved people from the devastating Boxing Day tsunami, caused by a vast undersea earthquake, by carrying them to higher ground. Studies in the USA have reported increased incidence of lost pet reports in local newspapers just before seismic activity.

So what are these animals picking up that our seismographs are missing? If we knew the answer to that, we would be better at predicting earthquakes ourselves.

See also THE EARTH 147, THE PLANETS 375

EASTER ISLAND

153. How did the Easter Islanders move their large Moai statues across the island?

When the Dutch explorer Jacob Roggeveen became the first European to land on the remote Easter Island in the southeastern Pacific Ocean, on Easter Sunday 1722, he and his crew were astounded to find the huge statues known as Moai. Later investigations have revealed 887 of these statues, which are up to 10 metres (33 ft) tall and weigh up to 86 tonnes. Carbon-dating indicates they were carved between the thirteenth and seventeenth centuries, generally out of volcanic rock from a crater known as Rano Raruka. But how did the islanders move these massive statues, sometimes to places many kilometres away? The local belief was that divine intervention enabled the statues themselves to walk to their destinations. Putting that to one side, there remain three main theories:

(i) The islanders used wooden rollers, dragging the statues across them with ropes.

(ii) They used wooden sledges, perhaps mounted on wooden rollers, to carry the statues.

(iii) They attached ropes around the necks and bodies of the statues and waddled them upright across the terrain, one step at a time.

The first two theories were originally discounted, as there were no trees growing on the island when Roggeveen landed, but later investigations of pollen deposits indicated that severe deforestation had taken place before 1650, and that there was probably no shortage of trees when the statues were made. All the same, attempts to move the statues by any of the suggested methods have been generally unconvincing.

154. How do we decipher the Rongorongo script found on stone and wood carvings from Easter Island?

In the nineteenth century, around two dozen wooden tablets were found on Easter Island bearing inscriptions in an unknown form of hieroglyphics. They were first mentioned in 1864 in an account by the French-born friar Eugène Eyraud, who was based in Chile and became the first European to live with the Easter Islanders. Eyraud mentioned hundreds of tablets, but by the time they came to be investigated by others in 1868, all but twenty-four had vanished.

Many of the symbols used on the surviving tablets look like people or animals, and alternate lines seem to be written upside down, but all attempts to decipher the tablets, or even to confirm that Rongorongo (which means 'lines for chanting') is a language, have failed. Its lack of similarity to any other known script suggests that it may have been one of the very few independent inventions of a writing system in the history of humankind.

Eyraud died of tuberculosis in 1868. He is thought by some to have brought the disease to Easter Island, where it wiped out around a quarter of the population.

Knowledge and timber shouldn't be much used till they are seasoned.

Oliver Wendell Holmes (1809–94)

155. What happened to the civilization that erected the Easter Island statues?

Even before the Europeans brought disease and the slave trade to Easter Island, the native population had undergone a severe decline. In the early seventeenth century, the population is estimated to have been about 15,000. A century later, it was down to under 3,000. Deforestation, possibly connected with the arrival of the Little Ice Age around 1650, internal wars and even cannibalism have been blamed for the population decrease, but no convincing evidence has been given for any of these having such a dramatic effect.

According to oral tradition, there were once two classes on the island known as the Long Ears (because of their extended ear lobes) and the Short Ears. The Long Ears are said to have enslaved the Short Ears and forced them to carve the Moai statues, but one night the Short Ears rose in revolt and killed all the Long Ears. That would account for the sudden cessation of statue-building, which left many Moai unfinished, but no evidence has been offered to support the legend.

ECONOMICS

Economists are notorious for not knowing things, or at least for not reaching precise conclusions. After eighty years and hundreds of books and papers on the subject of the Great Depression in the USA, for example, economists are still arguing about what its primary causes were. So to be kind to economists, I have decided to include just two examples of economic uncertainty.

156. What is the relationship between money supply and inflation?

Vast acreages of books and papers have been written on this, but basically it all comes down to the 'equation of exchange' proposed by the American economist Irving Fisher in 1911:

$$MV = PT$$

Where M is the amount of money, V is the velocity of circulation, P is price and T is transactions.

To see what this means, and why it's likely to be true, we need to look at what each means in a little more detail. If we take P to be a measure, such as the retail price index, of how much things cost, and T as the number of things people buy in a year, then PT will reflect the total amount they spend. On the left-hand side of the equation, M is the total amount of money sloshing around in the economy, and V is the number of times each unit of money is utilized in a financial transaction. So MV once again equals the total amount spent.

Now some people say that all the money in circulation is being used all the time, even if it's just being used to accumulate interest, and consumer demand for goods doesn't change much, so V and T remain more or less constant, which means that P, the measure of average prices, must be proportional to M, the money in circulation. Prices go up when the money supply is increased and fall when it is cut. QED on all counts.

Of course it's nowhere near as simple as that, as there are a handful of different definitions of money supply, according to whether they include credit and other notional borrowings and financings as well as printed banknotes. There is also a debate on whether to include values of investments and stock prices in the calculation of P, not to mention the continuing problems of arriving at a precise definition of V.

All the same, in the European and Japanese economies of the twentieth century, and also to a lesser extent in the US economy, there does seem to have been a correlation between money supply and interest rates, which are strongly linked to inflation. Since 1995, however, this correlation has been distinctly less convincing.

An economist is an expert who will know tomorrow why the things he predicted yesterday didn't happen today.

Lawrence J. Peter (1919–90)

157. Why are monkeys as irrational as humans in their economic behaviour?

Since 2005, research at Harvard and Yale Universities has demonstrated surprising similarities between the economic decisions taken by monkeys and people. The research began with experiments on tamarin and capuchin monkeys in which they were given tokens they could exchange for food. It quickly became clear that the concept of 'money' was something the monkeys could pick up without difficulty, and once this was established, their reactions to economic change could be tested.

In one early experiment, the monkeys were 'sold' portions of different types of vegetables and fruit at various prices, and when their shopping behaviour had settled, the prices were effectively changed by doubling the size of apple portions that could be purchased for one token. Under such circumstances, economic theory would predict that more apples would be sold and less money spent on, for example, grapes and melon. Not only did the monkeys follow this general behaviour, but their change in shopping patterns was found to be within 1 per cent of the change predicted by economic theory.

That set the stage for an even more interesting set of experiments to see how monkeys behave in a situation in which human behaviour is irrational. The human version is based on two games:

In Game One you are given $100 and told that you must make a choice. Your option is either to toss a coin and be given an extra $100 if it lands heads, but nothing if it is tails, or you can dispense with the coin-tossing and just be given another $50.

In the first case, you end up with $200 or $100, with equal probability; in the second, you end up with $150, come what may. So both are worth $150 on average and, in pure economic terms, there is nothing to choose between them. Yet when people are offered this choice, most of them are found to take the extra $50 and settle for a certain $150 rather than take the risk of ending up with only $100.

Game Two is essentially the same, but presented in a different way: you start with $200, but now the coin toss will either leave your money intact or lose you $100, while the other option is to hand back $50 and dispense with the coin-tossing. Again, you end up with $200 or $100 in the first option or a certain $150 in the second. But in this case, most people opt for the coin toss, apparently at least to give themselves a chance of hanging on to all their money.

Since the two games are mathematically identical, the differing behaviour by humans is not easy to explain. Interestingly, however, capuchin monkeys have been found to show precisely the same irrationality. In their case, there was no coin-tossing, but an experiment was designed with grapes to produce an analogous pair of games involving making a choice between which of two salesmen they bought grapes from.

In the first experiment, each salesman offered a plate containing two grapes. One always delivered both grapes when the money token was handed over; the other added one grape half of the time, but the other half of the time took one away. The monkeys preferred to do business with salesman number one.

In the second experiment, the salesmen both offered three grapes. Salesman one, however, always removed one of the grapes before handing the plate over; salesman two sometimes handed over all three grapes, but sometimes removed two of them. This time, the monkeys went for salesman two.

Whether this is evidence of a type of mental irrationality that monkeys and humans have shared since the two lines of primate began to diverge some 350 million years ago, or whether there are different reasons behind similar irrationalities is unclear.

Real knowledge is to know the extent of one's ignorance.

Confucius (551–479 BC)

EGYPTOLOGY

158. Were human sacrifices made at royal tombs in ancient Egypt?

Excavating the 5,000-year-old tomb of Pharaoh Aha-Mena in 1899–1900, the pioneering British archaeologist Flinders Petrie reported the discovery of an area to the east of the royal tombs he called the Great Cemetery of the Domestics. This consisted of thirty-four smaller graves, which he believed contained the remains of the pharaoh's servants. One theory is that they were put to death by poisoning at the time of the pharaoh's funeral, so that they could continue to serve him in the afterlife.

There is also some evidence that criminals or others may have been put to death during funerals as a sacrifice to the gods. While both these theories are consistent with an interpretation of inscriptions and also with some human remains found in the tombs, other explanations are also possible.

159. Where was the Land of Punt and where is the lost capital of Itj-Tawi?

During the Fifth Dynasty, around 2500 BC, Pharaoh Sahure organized an expedition to the Land of Punt, which was apparently very rich in commodities such as gold and ivory that were highly prized in Egypt. Several more expeditions to Punt are well documented over the next thousand years, but its location is never made clear. Some historians place it in East Africa, others in Arabia.

Another lost location is the capital city of Amenemhat I who was the first pharaoh of the Twelfth Dynasty, ruling from 1991 to 1962 BC. It is said that when he came to power he founded a new capital, which he named Itj-Tawi, 'the one that seizes the two lands'. It is thought to have been near the village of el-Lisht, where he built his pyramid, but no trace of this capital city has ever been found.

160. Was the labyrinth at Heracleopolis a myth or a reality?

Heracleopolis is the Greek name of Henen-nesut, which was the capital of Lower Egypt. According to the Roman writer Pliny the Elder, there was a vast labyrinth at Heracleopolis, said to contain forty shrines and many pyramids and temples to all the Egyptian gods. In 1940 a British archaeological team was reported to have discovered the labyrinth, but one of the team members fell ill and another disappeared, after which the excavation had to be abandoned. The location of the labyrinth remains a mystery – if it indeed existed in the first place.

161. What happened to Nefertiti?

Nefertiti, whose bust in the Egyptian Museum in Berlin may be the most copied ancient Egyptian relic after the mask of Tutankhamen, was the 'Great Royal Wife' (i.e. the principal consort) of Akhenaten, who ruled from around 1379 to 1362 BC. However, after about fourteen years of his reign, all references to her vanish. She may have died of plague, which was sweeping through Egypt at the time, or she may have ruled Egypt as Neferneferuaten after Akhenaten's death. More than one mummy has been identified with Nefertiti, but in all cases the evidence is very weak.

Knowledge comes, but wisdom lingers.

Alfred, Lord Tennyson (1809–92)

162. How did Tutankhamen die and who were his parents?

In 1922 the English archaeologist Howard Carter and his sponsor Lord Caernarvon made an astonishing discovery in Egypt's Valley of the Kings: it was not just the tomb of an Egyptian pharaoh, but uniquely one that had escaped the attentions of grave-robbers. The pharaoh in question was Tutankhamen, who ruled from around 1333 to 1323 BC, and thanks to his glorious funeral mask and the other beautiful artefacts found in his tomb, Tutankhamen has become the most famous of all ancient Egyptians. Yet until recently we knew almost nothing about him. He was about 1.675 metres (5 ft 7 in) tall, had large front teeth with an overbite, came to the throne at the age of nine or ten, and died at eighteen.

A computerized tomography scan in 2005 showed that he had broken his leg shortly before his death and that it had become infected, while DNA analysis in 2010 showed traces of malaria in his body. Either of those would have been enough to account for his early death, though signs of injuries detected in X-rays of his mummy together with evidence of possible plotting against him have led to some speculation that he may have been assassinated.

The DNA analysis also confirmed that Tutankhamen's father was the mummy known as KV55 and his mother was KV35, known as 'the Younger Lady'. There is much support for the view that KV55 was the pharaoh Amenhotep IV, who adopted the name Akhenaten, and that KV35 was his wife and sister, whose name is unknown. However, the identification of these two mummies has been disputed. We do, however, know that Tutankhamen's wet nurse was called Maia.

163. Did Egyptians really wear perfumed cones of fat on their heads at special events?

In Egyptian art, women, and sometimes men, are frequently depicted with strange conical structures balanced on their heads.

The usual explanation is that these were cones of fat that had been impregnated with perfume. As the event progressed the fat would melt, so releasing its perfumes on the wig or hair of the wearer.

However, in the absence of evidence either of greasy wigs or the means of attaching these fat cones to the wearer's head, some have suggested that the depiction of these cone-headed people was merely an artistic device to suggest good cheer, nice smells and riches sufficient to buy perfume.

See also MEDICINE 287, MURDER 316, PYRAMIDS 396–8, SPHINX 440–4

EINSTEIN

164. Was Einstein right about the cosmological constant?

When Albert Einstein proposed his theory of general relativity in 1916, he was concerned that the mathematics behind it suggested that gravity would cause the universe to contract and eventually to collapse in on itself. At the time, Einstein shared the general belief that the universe was unchanging in size, so to make his theory fit that idea, he introduced the notion of a 'cosmological constant' (→ COSMOLOGY 108). This allocated a density and pressure to empty space, so reducing the effect of gravity.

In 1929 the American astronomer Edwin Hubble made careful measurements that showed that other galaxies are moving away from ours, and that the universe is not unchanging in size, but actually expanding. Faced with the evidence, Einstein described the cosmological constant as the 'biggest blunder' of his life, yet recent developments in cosmology suggest that he may have been right after all – for established ideas run into problems when considering where the universe came from and where it is going. We may now be agreed that everything started with a Big Bang, but we cannot explain what held the universe together for long enough after that bang for matter to form and eventually coalesce into the stars and galaxies we now know. We can also calculate how old the universe is, based on the distance light has travelled since then. But we cannot explain why similar calculations seem to make some stars older than the universe itself. Einstein's cosmological constant may explain both these anomalies, and experiments now in progress at the Large Hadron Collider (LHC) at CERN in Switzerland may confirm that he was right after all. Einstein's 'biggest blunder' may yet come to the rescue of cosmologists.

165. If time and space are as inextricably linked as Einstein seemed to demonstrate, why is time so different from the other dimensions?

Einstein's idea of the space-time continuum is one of the most powerful and profound scientific ideas of all time. Writing of the unsustainability in his new physics of the concept of 'now', he said: 'It appears therefore more natural to think of physical reality as a four-dimensional existence, instead of, as hitherto, the evolution of a three-dimensional existence.' Einstein's theory of relativity had shown that time itself moves at different rates for different observers, and this had made it necessary to see the world as embedded in a four-dimensional space-time continuum, rather than three dimensions moving forwards in time.

But if time is on a some sort of a par with the other dimensions, why does it seem so different, and why can't we travel backwards in time?

Einstein later wrote that 'the separation between past, present and future is only an illusion, although a convincing one' – which is why this is such a difficult concept for us poor time-bound creatures to come to terms with.

166. What were the jokes that Einstein told his parrot?

In 2004 a sixty-two-page diary was found at Princeton University, covering the period from October 1953 up to Einstein's death in April 1956. It had belonged to Johanna Fantova, who was a close friend of Einstein in his final years. As well as recording Einstein's thoughts on physics and the politics of the time, Fantova recounts how, on his seventy-fifth birthday, Einstein was given a parrot. The great man decided the bird was depressed, and tried to cheer it up by telling it bad jokes. But what those jokes were, the parrot never divulged.

> Education is a progressive discovery
> of our own ignorance.
>
> Will Durant (1885–1981)

167. What were Einstein's last words?

Einstein died in hospital at Princeton, New Jersey, on 18 April 1955. Before he passed away, he uttered a few sentences in German. Unfortunately, the nurse attending him at the time did not speak the language, so his last words are lost. He did, however, scrawl down a few words: 'Political passions, once they have been fanned into flame, exact their victims…' The writing then trailed away.

See also BLACK HOLES 50, COSMOLOGY, 107–8, 110–11

ENGLISH HISTORY

168. Did King Arthur ever exist?

The earliest known mention of the legendary British leader known as Arthur dates back to the *Historia Brittonum* ('history of the Britons'), written in the ninth century, probably by a Welsh monk known as Nennius. Arthur only merits one paragraph in this chronicle, which does, however, list the twelve great battles he is supposed to have fought in the fifth or sixth century against the Saxon invaders. Details of these battles, such as Mons Badonicus (in which the Britons decisively defeated the Saxons at an unknown location), tie in with other historical accounts of the time. However, the earliest account of Mons Badonicus, that written by the sixth-century monk Gildas in his *De Excidiu et Conquestu Britanniae* ('on the ruin and conquest of Britain'), does not mention Arthur.

Arthur makes some brief appearances in accounts dating from the tenth and eleventh centuries, but it was the *Historia Regum Britanniae* ('history of the kings of Britain') by the twelfth-century Welsh cleric Geoffrey of Monmouth that fleshed out the story, providing the earliest known mentions of Merlin, Guinevere, Lancelot and the sword Excalibur. The earlier references had only identified Arthur as *dux bellorum* – a sort of warlord or military commander – but Geoffrey of Monmouth made him a king.

Probably inspired by Geoffrey of Monmouth, French literary romances of the twelfth and thirteenth centuries embellished the story even further, adding, for example, Arthur's legendary castle at Camelot. By the time Thomas Malory wrote *Le Morte d'Arthur* in the late fifteenth century, the Arthurian legend in its familiar form, complete with knights and Round Table, was firmly established. How much any of this tallies with the original Arthur, or whether Geoffrey of Monmouth made it all up rather than relying on now lost historical sources as he claimed, is a matter of pure speculation.

169. Why did the seventh-century Anglo-Saxons bury a longboat at Sutton Hoo, which was found with no human remains inside?

In 1939 a longboat 27 metres (almost 90 ft) in length was found buried at Sutton Hoo in Suffolk. It was one of the most magnificent and perplexing archaeological finds ever made in Britain. Inside the boat the archaeologists found golden artefacts and other precious treasures, but no human remains. The site was known as a cemetery, dated to the sixth and early seventh centuries, but this burial of a treasure-filled ship was unlike anything else. Two theories have been put forward to account for the absence of a body: either a body was originally buried with the ship, but it had been completely dissolved by chemicals in the soil; or the tomb was an early cenotaph to commemorate a powerful dignitary of the time. In either case, the person concerned is thought to have been Raedwald, a powerful East Anglian king of the seventh century. But whether he was buried with the ship or not may also be a secret that has dissolved into the soil.

170. When Edward the Confessor died in 1066, did he really say to Harold Godwinson 'I commend my wife to your care and with her my whole kingdom'? If so, what did he mean by it?

On his deathbed in the early days of 1066, King Edward the Confessor is alleged to have said those words to Harold Godwinson, apparently in contravention of his earlier promise to make William of Normandy his heir. Harold was promptly crowned king of England, but nine months later lost his life at the Battle of Hastings.

Did King Harold and his supporters invent Edward's supposed deathbed words? The earliest source appears to be the *Vita Ædwardi Regis* ('life of King Edward'), completed around 1067 and commissioned by Edward's wife, Edith of Wessex (whose marriage to Edward the Confessor was childless). Edith, the daughter of Godwin, Earl of Wessex, was Harold's sister.

Or did Edward in fact speak those words, intending only that Harold should look after the kingdom until William arrived to accede to the throne? As Edward's final illness was described as a 'malady of the brain' (most probably either a brain haemorrhage or some other form of stroke), we might also ask whether Edward was coherent enough to make his wishes clear.

171. What was the fate of Hereward the Wake?

Hereward the Wake ('the watchful one') was an Anglo-Saxon warrior who in 1070 led a revolt in the fens of East Anglia against the Norman takeover of England. He seems to have had a rebellious streak from his youth, having been exiled at the age of eighteen for disobeying his father. The anti-Norman revolt collapsed following the capture in 1071 of Hereward's base on the Isle of Ely. Hereward escaped by water, but what happened to him subsequently is more a matter of legend than fact. Some say his resistance continued and he was killed by Norman knights; others hold that he was pardoned by William the Conqueror and lived the rest of his life in peace; and the final possibility is that he never received a pardon but disappeared once more into exile and obscurity. The romanticization of his life and legend in Charles Kingsley's 1865 novel *Hereward* turned him into an archetypal English hero for the Victorian schoolroom.

172. Was the bricklayer Richard of Eastwell the illegitimate son of Richard III?

Richard III is known to have had at least two illegitimate children. There was John of Gloucester, who was appointed Captain of Calais in 1485 by his father as 'our dear bastard son', and there was Katherine Plantagenet, whose marriage in 1482 to the Earl of Pembroke was financed by the king. But there is a great mystery surrounding a third alleged illegitimate offspring, Richard of Eastwell.

Apart from a burial record for 'Rychard Plantagenet' in the parish records at Eastwell, Kent, in 1550, there is no hard evidence in support of the claim, but a number of circumstantial factors give it some credibility.

In 1546 Sir Thomas Moyle, Speaker of the House of Commons, was having some building work carried out on his property. When the workmen took a break, he noticed that while the other workers chatted and drank, one old bricklayer sat on his own reading a book in Latin. Sir Thomas went to talk to the man and gained his confidence, after which the man told his story. He had, he said, lived with a Latin schoolmaster until he was fifteen or sixteen, and did not know who his parents were, but a distinguished gentleman visited four times a year and paid for his upkeep. At the age of sixteen he was taken by the gentleman to Bosworth Field, where the forces of Richard III faced those of the would-be usurper, Henry Tudor. The boy was introduced to the king, who told him he was his son, and that he would acknowledge him if he won the battle. If he lost, the king advised the boy to conceal his identity forever. Richard was killed, Henry took the throne, and the boy fled to London, where he was apprenticed to a bricklayer.

Impressed with the tale, Sir Thomas Moyle allowed the man to live on his estate until his death. This was the story, at any rate, included by the English antiquary Francis Peck in his *Desiderata Curiosa*, a two-volume miscellany published in 1732 and 1735.

173. Was Amy, the wife of Robert Dudley, Earl of Leicester, killed so that Dudley could marry Queen Elizabeth I?

Elizabeth I was in love with Robert Dudley; Dudley was married to Amy Robsart; Amy died from a broken neck at the age of twenty-eight, her body being found at the foot of a short flight of stairs at Cumnor Place, near Oxford in 1560. The official inquest returned a verdict of 'misfortune', but the rumour at the time was that the 'misfortune' had been engineered by William Cecil, adviser to the queen, or by Dudley himself. The tragedy was subsequently fictionalized by Sir Walter Scott in his novel *Kenilworth* (1821). In this, Amy is the victim of a complex plot of deceit by Leicester's villainous steward, Varney. At the end of the novel, while Leicester entertains Elizabeth at Kenilworth, Varney engineers Amy's death at Cumnor Place, where she falls through a trapdoor.

In the mid-1950s a theory was proposed that Amy Dudley suffered from breast cancer, which could have weakened her spine, causing it to snap after a short fall. More recently, the rediscovery of the findings of the inquest has shown that the evidence could support either a fall or a more violent end.

174. Was the Gunpowder Plot an elaborate government conspiracy?

When Guy Fawkes was stopped, in the nick of time, from blowing up the Houses of Parliament in 1605, the case against the plotters looked clear-cut. It was without doubt a revolt by English Catholics against the anti-Catholic measures of King James I, whom they planned to assassinate, possibly with the idea of installing his daughter, Princess Elizabeth, as a Catholic queen (although she was not a Catholic).

That is all, no doubt, true. But there are several elements of the plot and its foiling that have suggested to many that there may be more to the story – and that Guy Fawkes and his colleagues may have been set up in an elaborate government sting, probably concocted by Robert Cecil, the king's ferociously anti-Catholic chief minister. Much of this conspiracy theory centres on a letter received by Lord Monteagle warning him to stay away from the House on the day the explosion was planned. The letter came from his cousin Francis Tresham, one of the plotters, who, it is said, may have been a double agent working directly for Cecil. Cecil's motive, according to this theory, was simply to discredit the Catholics in as dramatic a manner as possible. Otherwise, the theorists ask, how were the plotters able to rent a house so near to Parliament, how did they acquire so much gunpowder with such apparent ease, and – most of all – how did they manage to smuggle it into the cellar of the House of Lords?

See also ANCIENT HISTORY 14, BOUDICCA 54, DISEASE 119, THE MIDDLE AGES 298–300, 304, MURDER 319–19, 321, 324, SEX 421

175. Was James de la Cloche the illegitimate son of Charles II, as he claimed?

In 1862 the British historian Lord Acton received documents from Jesuit archives in Rome telling the extraordinary tale of James de la Cloche. The story begins in 1646, when the future King Charles II of England, then only in his teens, visited Jersey. Here he had an affair with Lady Marguerite de Carteret, as a result of which he fathered a son known as James de la Cloche.

The documents include three letters from Charles, which acknowledge that he was the father and relate how he granted James an annuity of £500 a year as long as he remained in London and practised the Protestant faith. James, however, went to Rome and joined a Jesuit seminary. He may even have acted as the king's secret and unofficial ambassador to the Vatican when Charles was contemplating converting to Rome (which, in fact, he only did on his deathbed, in 1685).

All mention of James de la Cloche in Rome ceases abruptly in 1668, but the following year a man calling himself James Stuart arrived in Naples, with the story that he was the illegitimate son of Charles II. He died in 1669, and in his will included a specific request to Charles II to grant his son a principality.

Whether James de la Cloche and James Stuart were the same person, whether either was the natural son of Charles II, and whether the letters de la Cloche showed to the Jesuits were forgeries are all still unknown. If the whole thing was fabricated, it must rank James de la Cloche as one of the greatest confidence tricksters in history.

ENGLISH LANGUAGE

The English language has its roots in Anglo-Saxon (known to scholars as 'Old English'), which emerged from the languages of the Germanic invaders who began to take over the country after the Romans left in the fifth century AD. But English is far from a 'pure' language, and is full of words that can be traced to Latin, Greek and Norman French, and also a great range of other languages, from Arabic ('alcohol', 'algebra', 'algorithm') and Hindi ('pyjamas', 'jodhpurs', 'bungalow') to Inuit ('anorak', 'kayak') and Australian Aboriginal ('boomerang', 'kangaroo'). Etymological research has usually been able to track the origins of any given word. The older ones can be linked by sound or spelling to the languages they came from, while the newer ones can be tracked back to their first appearances in print. Yet there are 1,098 words listed in the *Oxford English Dictionary* as 'origin unknown', and another 206 with 'origin obscure'. The examples that follow scarcely scratch the surface of our linguistic ignorance.

176. Why did jazz musicians in the 1920s start referring to an engagement as a 'gig' – and why is their music called 'jazz' anyway?

'Gig' is such a recent addition to our language that you would have thought someone would know how it started, yet the *Oxford English Dictionary* simply says 'origin unknown', before listing its earliest known uses, from 1926 and 1927, both in *Melody Maker*.

Another 'origin unknown' word is 'jazz' itself, which the *OED* suggests may be related to the word 'jasm', which is an alternative to 'jism', meaning energy or vitality. But what about the origin of those words? The *OED* again just says 'unknown', but they have been linked to Mandinki, a language of West Africa, the ancestral home of many African-Americans: in Mandinki the word *jasi* means to act in a bizarre or unusual manner, and more generally, along the West African coast, *jazz* meant 'hurry up', and the word was adopted into Louisiana Creole with this meaning, whence (some propose) it came to be applied to the fast, syncopated music popular in New Orleans at the end of the nineteenth century. This 'jazz' music made its first appearance in New York in 1915, introduced by Freddie Keppard's Original Creole Band, and made a more lasting impact with the arrival in 1917 of Nick LaRocca's Original Dixieland Jazz Band.

177. Where did the word 'bloke' come from as a slang term for a man?

The word 'bloke' has been used to mean a man or fellow for at least 150 years, and in the earlier years of the twentieth century it was used specifically to refer to a ship's captain. The nearest the *Oxford English Dictionary* can come to suggesting an origin is to cite an unlikely suggestion that it may somehow be connected to a Romany (and originally Hindi) term for a man: *loke*.

178. Where did the word 'posh' come from?

It has long been claimed that the word 'posh' derives from an acronym printed on the more expensive tickets issued by P&O in the glory days of the great ocean-going liners. 'Posh', it is claimed, stands for 'Port Out, Starboard Home', indicating the side of the ship on which the wealthier passengers would be allocated a cabin on the outward and return journeys in order to get the best of the sun (if sailing across the North Atlantic from Britain to America), or in order to be shaded from the fierce heat of the sun (if sailing from Britain to India, in the days of the British Raj).

The trouble is that nobody has ever produced one of these fabled tickets, and until one turns up, lexicographers will remain highly sceptical. Other theories include a borrowing from the Romany word *posh*, meaning a half, which was used for a halfpenny, then to money in general, whence it could have come to denote anyone rich. Or it may simply come from a quick and lazy way to say 'polished'. Finally, some claim it comes from a character called Murray Posh in George and Weedon Grossmith's comic novel *The Diary of a Nobody* (1888–9), who is described as 'quite a swell'.

179. What was the origin of the phrase 'rule of thumb'?

Did carpenters in the distant past use the length of the top joint of their thumb as a measure, equal approximately to one inch? If so, the word 'rule' in the phrase 'rule of thumb' refers to the use of the thumb as a ruler. Or does the phrase refer to the practice among brewers of using their thumbs to test the temperature of the fermenting liquid? Or does the phrase refer to a medieval by-law, which permitted men to beat their wives, but only with a stick no thicker than their thumbs? The latter explanation is very popular among amateur lexicographers, but there no evidence at all to support it. In fact, that explanation was offered first in the 1970s, while the phrase has been in use at least since the 1690s.

180. What was the origin of the phrase 'the full Monty'?

For an expression that has only come into common use comparatively recently – especially after the release of the film *The Full Monty* in 1997 – the origin of the phrase is curiously obscure. Some say it is simply a corruption of 'the full amount'. Others claim that it is a reference to the hero of El Alamein, Field Marshal Bernard Montgomery (1887–1976), affectionately known to his men as 'Monty', and that the phrase refers either to his full chest of medals, or to his liking for a full English breakfast. Others again suggest that the phrase might be a reference to full three-piece suits sold by the tailor Montague Burton (1885–1952), or to bets placed at the famous casino in Monte Carlo, or to full bales of wool from Montevideo.

A combination of the field marshal and the tailor seems the most likely, though who came first is anyone's guess.

181. How did the game of golf get its name?

There is a popular tale, for which there is no supporting evidence whatsoever, that the first golf club displayed a sign saying 'Gentlemen Only, Ladies Forbidden', giving rise to an acronym by which the game became known. That is almost certainly entirely apocryphal. The earliest known appearance of the word 'golf' was in an act of King James II of Scotland in 1457 banning the game, and this was centuries before people began to form words from initial letters. Indeed, the earliest citation for the word 'acronym' in this sense in the *Oxford English Dictionary* dates back only to 1943. Perhaps the 'Gentlemen Only' sign was advanced as a derivation because nobody seems to have come up with any other convincing explanation. The Scots word *gowf* means a blow or slap, but it seems more likely that the word in this sense came from the game rather than the other way round. There is also a Dutch word *kolf*, meaning a club or bat, which may have something to do with it, but no link has ever been established.

While we are on the subject of false acronyms, perhaps I should also mention that there is no evidence that those accused of certain immoral acts ever had 'For Unlawful Carnal Knowledge' stamped on their charge sheets.

EVOLUTION

182. How did life on Earth begin?

Our planet has been in existence for about 4.6 billion years, and for about the last 3.5 billion years there has been some sort of life on it. Even the simplest life forms that have been detected, however, are remarkably complex organisms, and the idea that they could have arisen by chance arouses very different reactions, even among scientists. Some say there's no real problem, as there are hundreds of billions of galaxies, each with hundreds of billions of stars, so even events that seem impossibly unlikely are still quite likely to have happened somewhere in the universe.

Others insist that the simplest life forms must themselves have evolved from simpler self-replicating molecules (such as DNA), and there is currently a good deal of research aimed at creating such organisms in the laboratory. This could provide evidence that such events could have happened in the conditions on the early Earth.

183. How did complex cells evolve from simple ones?

In 1952 Stanley Miller and Harold Urey at the University of Chicago performed an experiment showing that amino acids, which have been called the 'building blocks of life', could have formed spontaneously from inorganic chemicals thought to have been present in the Earth's atmosphere almost 4 billion years ago. While essential to life, however, amino acids cannot themselves be considered life forms. They are the components of proteins, a wide range of even more complex molecules which – along with nucleic acids (such as DNA), carbohydrates and lipids – are essential to the structure and function of the living cell as we know it today. But what form the first truly living things took is still a matter of speculation.

The most likely candidates are prokaryotes, simple single-celled organisms such as bacteria, which do not have a nucleus. Their name comes from the Greek karyon, meaning 'kernel', and pro-, 'before' or 'prior'. The name distinguishes them from the far more complex eukaryotic (the prefix eu- meaning 'true' or 'good') cells, as found in the group of more evolved unicellular organisms known as protists (including, for example, amoebae and single-celled algae), together with plants, fungi and animals. Although both types of cell are DNA-based, their structures and mode of functioning are very different, which makes it difficult to see how eukaryotic cells could have evolved from prokaryotic ancestors. The most plausible suggestion is that groups of prokaryotic cells may have joined together in symbiotic (mutually supportive) colonies, from which the eukaryotic cells developed. The process may even have started with one cell eating another. Yet it is unlikely that we will ever be able to produce any incontrovertible evidence that such a process actually took place.

184. Which came first, the gene or the protein?

The very basis of life as we know it consists of amino acids organized into proteins, whose construction is mediated by the nucleic acids DNA or RNA. But the nucleic acids are themselves synthesized through processes involving proteins. So proteins and DNA (which carries genetic information) appear to be inseparable, yet they are too distinct to have evolved together.

One theory, known as the RNA-world hypothesis, is that life on Earth began with organisms based on RNA (ribonucleic acid), from which DNA and genes evolved, which gave rise to proteins. In contrast, the iron-sulphur-world theory proposes that the earliest life comprised a very simple form of metabolism, without genetics, that occurred on the surface of minerals through complex chemical reactions fuelled by volcanic activity. These early cells could then have evolved into proteins, from which the nucleic acids and genes emerged.

185. Who or what was LUCA (the Last Universal Common Ancestor)?

Just as the name 'Mitochondrial Eve' has been given to the most recent woman whom all people on Earth can claim as a direct ancestor, the acronym LUCA has been bestowed on the hypothetical living organism much further back in time from which all organisms now alive descended. Since the genetic code was first deciphered in the 1960s, we have come to realize that all cellular life has a great deal in common. All life stores its genetic information on DNA, in packets called genes, which contain the recipes for making RNA and proteins.

If everything living operates under the same rules, it seems reasonable to assume that way back it all stemmed from the same common ancestor, and the last such ancestor must have been around after DNA had evolved, but long before life forms split into the myriad variants we see today, from bacteria to plants, reptiles, mammals and humans.

186. Has evolution in general been a gradual process, or have species evolved by sudden jumps between periods of stasis?

There are two theories about the pace at which the evolutionary process works. Darwin's own conception was that it is gradual and going on all the time; small changes in organisms build up over long periods into big ones. One objection to that idea, however, was the absence in the fossil record of evidence of the intermediate forms that would substantiate this gradualist theory. Darwin himself was troubled by these evolutionary missing links, though others have argued that the fossil record should be expected to have a great deal missing, as fossilization is something that only happens by chance, when conditions are appropriate.

In 1972, however, the American palaeontologists Stephen Jay Gould and Niles Eldredge proposed a theory they called 'punctuated equilibrium', which proposed that species tend to remain the same for long periods, then change with a sudden jerk. Furthermore, Gould and Eldredge maintained that such changes do not occur in the mainstream population, where interbreeding is likely to nullify rather than encourage changes, but on the fringes of a population, where a group with the modified gene live in relative isolation, allowing the genetic change to be preserved, if it is better suited to survival than the older genetic make-up. While evidence has been found in the fossil record of some examples that appear to conform to the patterns of punctuated equilibrium, it is by no means clear whether such patterns are normal or only of rare occurrence. Those who believe such patterns are rare hold to the theory called 'punctuated gradualism', which maintains that generally speaking the process of evolution is slow and gradual, but every now and again it speeds up rapidly.

See also DNA, GENETICS 206–14, GIRAFFES 215–16, HUMAN EVOLUTION 237–41

FOOTBALL (AMERICAN)

187. To what extent are American footballers and dwarfs prone to chronic traumatic encephalopathy (CTE)?

CTE is a neurodegenerative disease caused by repeated trauma to the head. However, its symptoms – which may include tremors, speech problems and dementia – usually occur only many years after the injuries. Signs of CTE were first observed in boxers about eighty years ago, and was associated with being 'punch drunk', but more recently it has been observed in American footballers – and in a dwarf who had participated in dwarf-throwing contests, in which he had been knocked out on a dozen occasions.

Yet nobody knows why symptoms take so long to appear, or how many hits to the head, or how hard those hits need to be, for a person to develop CTE. Nor is it known if there are other causes besides blows to the head.

There are many things of which a wise man might wish to be ignorant.

Ralph Waldo Emerson (1803–82)

FRENCH HISTORY

188. Who was the Man in the Iron Mask?

From his arrest under the name Eustache Dauger in 1669 until his death in 1703 (when he was known as Marchioly), the identity of this prisoner of Louis XIV was kept a secret. Nobody was even allowed to see his face, which was kept hidden behind an iron mask. It has even been suggested that he was James de la Cloche, the alleged illegitimate son of Charles II (→ ENGLISH HISTORY 175). Did all who knew who he was take the secret with them to their graves, neither writing it down nor telling anyone? Over three hundred years later, we seem no nearer solving the mystery. Films have been made and books written in which a number of candidates have been put forward, but no real evidence has been found.

Voltaire suggested that the man in the iron mask was an older, illegitimate brother of Louis XIV; the historian Hugh Ross Williamson thought he might have been the king's natural father; but such claims that the prisoner had royal blood have been disputed on the grounds that he applied to work as valet to the prison governor, which would have been a very un-royal task. Others have proposed General Vivien de Bulonde, who, when fighting the Austrians, incurred the king's wrath by ordering a hasty withdrawal of the troops under his command. But given the prisoner was prevented from talking to or meeting anyone other than one guard at the jail, and that all his clothes and belongings were destroyed when he died, only circumstantial evidence has been offered for any of these candidates – who represent only a handful of those who have been suggested over the centuries.

Perhaps somewhere, in an old library in France, there is a forgotten slip of paper containing the answer to this mystery. The vital clue may not exist or may never be found, but it could turn up tomorrow.

189. Did anyone ever say 'Qu'ils mangent de la brioche'?

That line – so often attributed to Marie Antoinette, wife of Louis XVI, and most commonly translated as 'Let them eat cake' – seems to have its origins in Jean-Jacques Rousseau's *Confessions*, which were written in the late 1760s, when Marie Antoinette was still a teenaged Austrian archduchess living in Vienna. Rousseau says they were spoken by 'a great princess' on hearing that the peasants had no bread, but he does not specify who this princess was.

Louis XVIII, in his memoirs, describes the line as an old legend, and says that within his family it was usually attributed to Maria Theresa, wife of Louis XIV. Those memoirs, however, were written after Rousseau's *Confessions*, which may have influenced his thoughts.

The connection with Marie Antoinette no doubt came about through her growing unpopularity in the years leading up to the French Revolution, a time when her opponents would gladly have seized upon Rousseau's story and attached her name to it. But where Rousseau got it from is not known.

190. What colour was Charlotte Corday's hair?

Charlotte Corday, who was guillotined in 1793 (→ CONSCIOUSNESS 104) for the assassination of the radical Jacobin leader Jean-Paul Marat, became a martyr and heroine of the Girondists, the moderate party during the French Revolution. But what colour was her hair?

Of the many paintings of her, most were done posthumously and portray her as dark-haired, but in some portraits done from life, she is depicted as a brunette. Her passport, however, describes her hair colour as chestnut, while a crime scene painting done by Jean-Jacques Hauer shortly after the murder shows her as fair-haired. It has been suggested, however, that for the purposes of that painting, Hauer wished to portray her as vain and aristocratic, so made her look as though she had powdered her hair.

191. Who was the first person to cross the Pont Julien?

The Pont Julien near Lacoste, a picturesque mountain village tucked away in the Vaucluse department of southeast France, is a rare example of a Roman bridge dating back to the 1st century BC. The bridge was closed to traffic in 2005, when the last person to cross it was Finnbar Mac Eoin, author of *Two Suitcases and a Dog*. A plaque on the bridge marks the event, saying: 'We do not know who was the first person to cross, but an Irishman was the last.'

See also THE MIDDLE AGES 298–300, MURDER 320

FUNDAMENTAL PARTICLES

192. Is there any hope for the theory of super-symmetry?

Physicists have long been searching for a 'Theory of Everything' that will explain in a coherent way all the forces that hold the universe together. Despite the development of quantum mechanics and the theory of relativity, the search seemed to be getting no further, but one of the latest candidates is string theory, of which we shall have more to say later (→ PHYSICS 369, QUANTUM PHYSICS 401).

One of the predictions of string theory, however, is that at higher energy levels we should start to see evidence of a symmetry in the elementary particles. Known as super-symmetry (or SUSY for short), the theory gives every particle that transmits a force (a boson) a partner particle that makes up matter (a fermion), and vice versa. The trouble is that none of these partner-particles has ever been detected. Testing the theory of super-symmetry was one of the objectives of the Large Hadron Collider at CERN in Switzerland. However, none of the predicted particles has yet been detected, and even some former supporters of the theory are suggesting that it may have to be abandoned.

193. Are quarks made of even smaller particles?

The Greeks had a word for a fundamental particle: they called it an atom, which meant indivisible. In 1897 the English physicist J.J. Thompson discovered the electron, showing that atoms were divisible after all. Later, it was found that the electrons were in orbit round a tiny nucleus, constituting most of the mass of the atom and comprising particles called protons and neutrons. Then, in 1964, the American physicist Murray Gell-Mann proposed that protons and neutrons were made up of even smaller particles, which he dubbed quarks (borrowing a word from James Joyce's novel *Finnegans Wake*). There are six different types, or 'flavours', of quark, known by the rather picturesque names of up, down, top, bottom, strange and charm, none of which can exist on their own, but which combine in various ways to form other particles. But whether the search for smaller and smaller particles will stop with the quark, nobody can say.

194. Do protons decay into quarks, which could then form other fundamental particles?

As mentioned above (→ 192), for many decades the Holy Grail of physics research has been the search for a 'Theory of Everything', bringing together the four fundamental forces governing the behaviour of matter. These are electromagnetism, the strong force (which holds elementary particles together to form atoms and molecules), the weak force (which is responsible for radioactive decay), and gravity.

As the Theory of Everything seemed to grow ever more elusive, the search was narrowed to finding a 'Grand Unified Theory' (GUT) that would explain the first three of these forces, leaving out the problem of gravity, which is a very much weaker force than the other three, although it works over great distances. With the development of the quark-based 'Standard Model' of particle physics, several GUTs have been proposed and their predictions tested experimentally.

One such prediction concerns the transformation of protons into neutrons and vice versa. This has been observed experimentally in free neutrons and protons, which are not bound to other particles in an atomic nucleus, but the bound versions are highly stable. In fact, the theories predict a decay process with a half-life of about 10^{32} years. That's about the age of the universe, with another 22 zeroes added to the end. Experiments to detect signs of such a process taking place have yet to produce any evidence supporting it.

195. Does the Higgs boson exist?

The 'Standard Model' of particle physics was formulated in the 1970s to bring some sort of order to the vast number of fundamental particles that subatomic physicists had found it necessary to define. This Standard Model describes all particles in terms of matter and force. Matter is made of a class of particle called fermions, which may be either quarks or leptons, and force is accounted for by another class of particle: bosons.

Quarks and leptons each come in six forms, held together by four types of boson, all of which have been confirmed experimentally. The fifth type of boson, however, has so far evaded detection. This is the infamous Higgs boson, named after the English physicist Peter Higgs who first proposed its existence in the 1960s. If it exists, the Higgs boson is what gives particles their mass and is therefore the key to the transformation of energy into matter in the first fraction of a second after the Big Bang – which is why it has been called the 'God particle'. If the Higgs boson does not exist, physicists will need to come up with another theory involving a modification of the Standard Model.

Experiments at the Large Hadron Collider at CERN are intended to test the theory by detecting the existence of the Higgs boson. If it is found, however, Stephen Hawking will lose £50, following a bet he made with Professor Gordy Kane of Michigan University in 2000 that it does not exist.

196. If the Higgs boson does not exist, what is responsible for energy turning into matter?

In 2008 Stephen Hawking said: 'I think it will be much more exciting if we don't find the Higgs. That will show something is wrong, and we need to think again.' A good deal of thinking has clearly been going on, as there are at least seven different theories that could fill the gap if the Higgs boson is not found by the Large Hadron Collider. Needless to say, all of these explain some aspects of subatomic physics but leave others unexplained. The Higgs boson is still the best candidate to take part in a Grand Unified Theory.

197. Why do some particles have mass, while others have none?

According to current theory, photons, which are particles of light, have no mass, while other particles, such as electrons and quarks, do have mass. Quite what the difference is, and why there is a difference at all, is unknown. According to supporters of the Higgs boson theory, a 'Higgs field' permeates the entire universe and interacts with mass-less particles to give them mass. This is another problem that the Large Hadron Collider may throw light on. But whether that light can be transformed into mass is another matter.

See also QUANTUM PHYSICS 399–407

GAMES

198. How did the children's string game of cat's cradle spread to so many different cultures?

The game of cat's cradle, played by two people and involving the manipulation of loops of string held around the fingers, clearly has a long history. It has been found among diverse populations around the world, with several anthropologists reporting attempts to teach it to children in isolated communities to gain their trust and friendship, only to discover that they already knew it.

Everyone from the Inuit of the Arctic to tribes in sub-Saharan Africa seems to know cat's cradle, but whether it was invented independently by a number of cultures or has one origin from which it spread throughout the world is unknown. Even the origin of its name is disputed. Some say that it is a corruption of 'cratch-cradle' (i.e. manger-cradle), but the *Oxford English Dictionary* describes that as a 'guess' that is 'not founded on facts'. The earliest known use of the name in English dates from 1768, but the game itself is probably much older.

199. What was the fifth card in Wild Bill Hickok's fatal hand besides the aces and eights?

When Wild Bill Hickok was shot in the back of the head on 2 August 1876 while playing poker in Saloon No. 10 at Deadwood, South Dakota, it was widely reported that he was holding two pairs, of aces and eights. Those cards are even known as 'Dead Man's Hand', but there are five cards in a poker hand, so what was the other one?

The earliest known account of the hand came many years later from Ellis Pierce, the town barber, who was called to prepare the body for burial. He told of the aces and eights, but did not mention the fifth card. The modern No 10. Saloon, built in the image of the original where Wild Bill was shot, displays the hand on its wall – with the fifth card as the nine of diamonds. The Adams Museum, however, also displays the hand, with the queen of hearts as the fifth card, while the five of diamonds, jack of diamonds and queen of clubs have also been claimed as the fifth card.

There is, of course, the possibility that only four cards had been dealt and the deal was interrupted when Hickok was shot...

200. Is there a sudoku puzzle with sixteen numbers in the initial set-up with a unique solution?

A sizeable number of seventeen-digit sudokus have been found, but no valid sixteen-digit example has yet been discovered – but neither has any proof that such a puzzle cannot exist.

201. Can bacteria solve sudoku puzzles?

In an extraordinary experiment at the University of Tokyo in 2010, *E. coli* bacteria were shown to be able to solve simple sudoku-like puzzles. The experiment involved sixteen types of modified *E. coli* bacteria, each of which was given a distinct genetic identity according to the square it had been assigned in a four-by-four grid. The bacteria were also designed to be able to adopt one of four 'colours'. Some of the bacteria were allocated colours; others were left to determine their own.

By means of RNA, the bacteria could transmit their position and colour details to each other. Additional programming of the genes prevented the bacteria whose colour had not been assigned from adopting the same colour as any other in the same row, column or two-by-two block as itself. In other words, the same rules as sudoku. The experiment showed not only that the bacteria choose the right colours to complete the grid, but that they 'solve' all the squares at the same time.

The researchers say that by expanding these principles to eighty-one types of bacteria, a full nine-by-nine grid could be solved. But that has yet to be confirmed in practice.

202. What is the result of a chess game if both sides play perfectly?

With best play, the game of noughts-and-crosses (tic-tac-toe in the USA) should end in a draw, as it is not difficult to demonstrate. In the more complex game of Connect Four, it was shown by exhaustive analysis in 1974 that the first player can force a win. The still more challenging game of gomoku (five in a row) was solved by computer in 1994, showing that the first player may again force a win with best play. But what about chess?

Results from international tournaments have supported the general view that White, who moves first, has an advantage, with results for White around 2 or 3 per cent ahead of Black. With around 10^{120} possible chess games (which is significantly more than the number of atoms in the universe, estimated at around 10^{80}), a complete analysis of chess is currently impossible. Whether White's advantage is enough to win, or whether Black can defend himself, or whether indeed with best play Black should actually win, at present appear to be unfathomable questions.

203. How many guesses does it take to win at Mastermind?

The game of Mastermind is a simple but tantalizing exercise in deduction in which one player places four pegs in a row from a set containing six colours and the other player, who cannot see the pegs, has to guess the colour of each of the concealed pegs. After each try, the guessing player is told how many pegs he has identified correctly and how many are the right colour but in the wrong place. The game continues until a correct guess is made.

Detailed analysis has shown that the guesser can always succeed in at most five moves, and that the average number required is 4.478. The unsolved question, however, is what happens if all the guesses have to be made right at the start, without waiting for the replies?

A set of six guesses is known that guarantee that the answers will allow the guesser to work out the other player's peg formation. If the colours are identified by the numbers 1, 2, 3, 4, 5, 6, just ask about (1, 2, 2, 1), (2, 3, 5, 4), (3, 3, 1, 1), (4, 5, 2, 4), (5, 6, 5, 6) and (6, 6, 4, 3) and the replies will let you work out the answer on your seventh try. However, no one has been able to work out whether this set of six can be reduced to five.

> The greatest enemy of knowledge is not ignorance, it is the illusion of knowledge.
>
> Stephen Hawking (b.1942)

GARLIC

204. Why are vampires supposed to be repelled by garlic?

The legend that garlic has the power of fending off vampires may have been popularized by Bram Stoker's novel *Dracula*, published in 1897, but it is much older than that. In Romania in particular, where much of Transylvania is now located, garlic has long been held as an effective way of warding off evil and curing disease. The practice of smearing corpses with garlic, or leaving cloves of garlic in their orifices in order to prevent evil spirits entering and to keep vampires away has also been recorded.

Some connect the garlic/vampire myth with the fact that mosquitoes are repelled by garlic. The idea that a measure to protect against one biting, disease-spreading creature was put into service against another is not totally convincing, however, as mosquitoes were only shown to spread various diseases (such as malaria and yellow fever) in the later nineteenth century. Whether the irritation of their bite alone was enough for them to be compared to Count Dracula is arguable. Garlic has long had the reputation of being health-giving, but no convincing explanation has been offered as to why it should be so repellent to vampires.

205. Does eating garlic reduce the rate of cancer and flu?

The belief that garlic is effective against a number of diseases goes back at least to the ancient Egyptians around 1500 BC. In the Middle Ages people used it in an attempt to ward off the plague, and it was given to soldiers in both World Wars in the belief that it prevented gangrene.

Over the past decade, much research has gone into the question of whether garlic has a beneficial effect against cancer, and a large number of studies have reported a positive result. Many of these, however, are based on population studies, comparing garlic consumption with cancer figures; others are based on laboratory tests on animals; and many are based on the injection of chemicals derived from garlic.

The jury is still out on the question of whether eating garlic has the desired effect, and if so, how much one ought to eat. Equally unproved by scientific studies is the reputation garlic vodka has in Russia as a way of preventing or treating flu.

GENETICS

206. How is the blueprint of the genetic code translated into building instructions through which a body is formed?

Thanks to the pioneering work of Crick and Watson (→ DNA) and the many who followed, we now know the alphabet of genetics and the details of what it spells out in the human genome. We know the genome comprises a vast string of nucleotides that come in four varieties known as A, C, T and G, which in various combinations provide the code of individual genes. We have identified the genes associated with certain characteristics and ailments, and we know that the same genes may be responsible for different but related parts in the construction of different organisms. But how does it all work? How is the genetic code read during the process of growth, and how do certain strings lead to the development of certain proteins, or blue eyes, or a susceptibility to a particular disease? We know the letters; we are beginning to pick up some of the words; but the whole process of how this is turned into a living creature is a mystery.

207. How can we predict how a gene will function?

Scientists have had great success in identifying the genes that are associated with certain functions or characteristics, but the reverse process is a very different matter. It's a bit like having a dictionary that translates English into Chinese, and trying to find a particular Chinese word in it. If that word is very similar to one that has been encountered before, you have grounds for making a good guess, but a radically new string of Chinese ideograms can pose impossible problems.

In the same way, if a string of nucleotides represented by the letters A, C, T and G is very similar to a known gene, then its function will probably be similar too. But when it bears only a small resemblance to anything seen before, we have no way of knowing what part of the body it controls or what its effect might be.

208. How do genes affect behaviour?

Long before we knew about the genetic code, and even before Darwin's theory of evolution, it was known that we inherit characteristics from our parents or more distant ancestors. Yet while the science of genetics has thrown much light on the inheritance of physical characteristics, the inheritance of behavioural characteristics has proven much more difficult to explain.

The long-running argument on the relative effects of nature and nurture on an individual's development has never been totally settled, but many studies on twins have shown that the development of attitudes, intelligence and certain personality factors all have at least a component of heredity in them. It may be that the balance of brain chemicals that can affect behaviour is directly controlled by genetic factors, but many behavioural factors are the result of a conscious decision-making process. The manner in which genes can affect such processes, or the extent to which they might do so, has yet to be discovered.

209. How do the various parts of an animal's body know when to stop growing?

Somehow, our cells (and those of every other living thing) read the genetic information from our DNA, transcribe it and copy it to messenger-RNA. The RNA is read by the next stage in the production line, which is the ribosome that uses the RNA information to produce sequences of amino acids, which in turn assemble into proteins. The RNA sequence even tells it when to stop. But the stopping instruction only refers to the end of an individual protein. How the machinery knows it has come to the end of a bone, or a nose, or a leg – and it is therefore time to stop – is another matter entirely.

210. Why don't identical twins suffer from the same genetic diseases?

Until recently, 'identical twins' were thought to have absolutely identical genes, so if one suffered from a disease that was known to be genetically transmitted, the other ought to have it too. If that was not the case, then the only available explanation was that environmental factors must be playing a part in causing the gene to do its worst, or to prevent it from so doing.

Recently, however, it has become increasingly clear that identical twins may not be genetically identical at all. A study of nineteen pairs of identical twins revealed variations in their DNA caused either by missing segments or segments that had been copied several times. The double-helix structure of DNA ensures a mechanism for repairing such errors, but for some reason it does not always operate, resulting in small differences.

Apart from copying errors, there is also something called 'gene over-expression', the situation when a gene produces more RNA and more protein than it ought to. Gene over-expression has also been identified in identical twins who do not share a genetic disease, leading to the suggestion that the over-expressed gene is responsible for causing the disease gene to come into action. But we do not know how or why.

211. Is there a genetic key that can switch off ageing in humans?

Genetic modification performed on certain living things has succeeded in doubling their lifespan or more. This has been successfully done with yeast, worms and mice. What causes the ageing factor in humans is not properly understood and a number of different factors are thought to be involved, but genes seem to play a significant part. In the organisms mentioned above, a factor was identified that turned age-regulated genes on and off. There is good reason to believe that something similar will one day be found in humans – although at this stage it is not possible even to guess the extent to which it may prolong life.

212. Does experience create a mechanism whereby genes may be turned on and off?

The idea of the French naturalist Jean-Baptiste Lamarck (1744–1829) that acquired characteristics can be inherited has been discredited since Darwin's day. But there is growing evidence for a sophisticated new theory called epigenetics, which brings back the idea of experience affecting aspects of our genetic make-up.

The basic idea behind epigenetics is not that experience can change our DNA, but that it may affect the activation of certain genes or of certain proteins associated with DNA. In some cases, the theory maintains, this may also result in the inheritance of whatever characteristic has been changed. Studies in different parts of the world have suggested that starvation, toxins and stress may all have genetic effects that last several generations.

213. How do bacteria swap genes?

Genes may pass from one generation of humans to the next, but recent discoveries have shown that bacteria have another way of doing it. Called 'horizontal gene transfer', it involves a bacterium picking up genetic information from near relatives, which may include genes that enable the bacteria to withstand antibiotics. This may help explain how bacteria are able to adapt so quickly to new environments, but how the transfer takes place has not yet been discovered.

214. Is the human genome inherently unstable and therefore bound to lead to our extinction?

The evolutionary history of the human genome has displayed its wonderful pliability, which has enabled us to change, adapt, survive and take control, to a large extent, of our environment. That same pliability, however, has fostered the parallel evolution of diseases and other life-threatening factors of ever-increasing sophistication and potential danger. A study of the extinction of other species over the ages shows that some have fallen prey to environmental change, natural disasters or similar external factors, but others have just died out, the seeds of their own extinction being apparently sown in their own genetic make-up. Will the human race go the same way?

See also AARDVARKS 1, DNA 132–4, EVOLUTION 182–6, HUMAN BEHAVIOUR 235, HUMAN EVOLUTION 237–41

GIRAFFES

215. Why do giraffes have long necks?

The old theory was that giraffes with long necks had an evolutionary advantage as they could reach the leaves that were higher in trees after the lower leaves had been eaten. There are at least three arguments against this theory, Firstly, when the long neck of the giraffe was evolving, the Earth was covered with lush vegetation and there would have been no leaf shortage at low levels. Secondly, giraffes are fussy eaters and go for certain types of leaf rather than the highest ones. Lastly, giraffes tend to eat with their necks held horizontally, not stretching upwards.

Alternative theories that have been suggested are that long necks evolved to improve a giraffe's ability to look for predators; that long necks evolved as a sexual signal; or that giraffes evolved long legs to help them run faster, and long necks then became necessary to reach to ground level to drink.

A detailed examination of all these theories, however, reached the conclusion that there was insufficient evidence to support any of them.

216. Why are so many male giraffes homosexual?

Several studies of animal sexual behaviour have remarked on the frequency of homosexual relations between male giraffes. One study even reported that over a period of observation of giraffe sexual behaviour, 94 per cent of activity was between two males, 5 per cent was male–female and 1 per cent female–female.

While some have pointed out the potential evolutionary advantage of a gay gene, in that it may provide additional male help in the family in looking after children, the gay giraffe figure is surprisingly high.

See also SLEEP 428

THE GREEKS

217. How did Socrates make a living?

Considering the huge influence of Socrates on the subsequent course of Western philosophy, it is remarkable how little we know about the fifth-century BC Athenian thinker. Almost all we do know comes from the philosophical writings of his followers Plato and Xenophon and the comic plays of Aristophanes. Nobody bothered with a biography. Socrates himself did not even write down his own ideas, for which we rely largely on Plato's *Dialogues*. Yet how much of these are Plato's own thoughts put into Socrates' mouth and how much was genuine Socratic wisdom has long been argued.

On the matter of what Socrates did for a living, we face not so much a lack of information as a good deal of evidence that he did nothing. In Plato's *Apology*, Socrates cites his poverty as evidence that he is not a teacher. Xenophon also has him denying that he ever accepted payment for his teachings and quotes him as saying that he devotes himself entirely to discussing philosophy. Later writers said that he followed his father's profession of stonemasonry, but no evidence has been found to support that assertion. Socrates did, however, marry the much younger Xanthippe, who bore him three sons, so he must have had some sort of income to support them.

What we do know is that he annoyed the Athenian establishment so much that in 399 BC he was tried and found guilty of atheism and corrupting the youth of the city. As a result he was sentenced to kill himself by drinking hemlock.

I know nothing except the fact of my ignorance.

Socrates

218. Did Pythagoras ever exist?

We may not know much about Socrates, but we know even less about the person who gave his name to the famous theorem about the square of the hypotenuse (→ THE UNIVERSE 461). The Pythagorean school of philosophy – which proposed that number and ratio underlay the workings of the universe – flourished as a secret sect in the late sixth century BC, but nothing was written about Pythagoras himself until hundreds of years later.

The lack of information may have been due, in part at least, to the secrecy of the brotherhood and the cult it spawned – whose followers believed, among other things, in the transmigration of souls and the sinfulness of eating beans. Some have suggested that there never was a man called Pythagoras, but that he was a fictional figurehead in whose name the theories and discoveries of the brotherhood could be issued.

219. What was the secret behind the Greeks' use of fire as a weapon?

Towards the end of the seventh century AD, the Byzantine empire, centred on the predominantly Greek-speaking city of

Constantinople, introduced a new secret weapon, used both on land and at sea. Described as 'liquid fire', it was projected from siphons on enemy positions or ships, bursting into flames on contact and being notoriously difficult to extinguish. Fire had long been used in battle, but this 'Greek fire', as it became known, was far fiercer than anything that had previously been seen, causing panic and destruction wherever it was used. The secret behind the weapon was passed from one Byzantine emperor to the next for centuries.

Even when siphons and samples of the flammable material and even entire fire ships were captured by enemies, attempts to reproduce the secret formula failed, and neither the principal chemicals in the composition of Greek fire nor the alleged 'secret ingredient' it contained were ever discovered. By the early thirteenth century the use of Greek fire as a weapon had faded away. Perhaps the formula had been lost, or perhaps its makers could no longer obtain the required ingredients.

See also ANCIENT HISTORY 17, 19, VENUS DE MILO 467

HAIR

220. Why do we have pubic hair?

There are several theories:

(i) Pubic hair evolved as a way of signalling our sexual maturity to members of the opposite sex.

(ii) It provides protection for a delicate part of our anatomies.

(iii) The hair helps the body retain the pheromones produced by our glands to signal to the opposite sex and entice them to reproduce. In other words, the hair is good because it makes us smellier.

221. Why does the hair on our heads grow so long?

From an evolutionary point of view, our ability to grow our hair so long seems strange. The common ancestor we share with the other apes is generally assumed to have been hairy. In the course of our evolution away from our ape relatives, we have gradually shed most of our body hair. Yet on our heads the hair grows much longer than that of any other ape, or indeed almost any other mammal. Most of the rest of the hair on our bodies grows to a certain length then falls out, which is the way most mammal hair behaves, yet on our heads, and even on men's chins, it may grow to a length of a metre or more.

Once again, sexual selection has been offered as an explanation, but whether it is the beauty of the long hair, or the belief that it signals some sort of fitness, or the smells it may retain, is not specified.

It is impossible to make people understand their ignorance; for it requires knowledge to perceive it and therefore he that can perceive it hath it not.

Jeremy Taylor (1613–67)

HANDEDNESS

222. Is it true that most, if not all, polar bears are left-handed?

Open almost any book of trivia and you will find the following claim: All polar bears are left-handed. You will also, in all probability, read that 'A duck's quack doesn't echo', which is complete nonsense, but has been copied by so many indiscriminate trivia peddlers from other dealers in junk factoids that the sentence now racks up 'about 226,000' hits on Google. By contrast, the left-handed polar bear statement scores only 'about 90,000', but the jury is still out on the matter.

The anthropologist Richard Nelson, who spent a year in the 1960s living in an Inuit (Eskimo) community in the village of Wainwright in the north of Alaska, just inside the Arctic Circle, reported the advice he had received to dive to your right if attacked by a charging polar bear:

> *Inupiaq elders say polar bears are left-handed, so you have a slightly better chance to avoid their right paw, which is slower and less accurate. I'm pleased to say I never had the chance for a field test. But in judging assertions like this, remember that Eskimos have had close contact with polar bears for several thousand years.*

It has been suggested that the story of the left-handed polar bear may have had its origins in an account by an Inupiaq elder who saw one polar bear using its left paw, from which he generalized. There is also an often repeated claim that polar bears, when hunting seals, cover their black noses with their right paws in order to conceal their presence against the snowy background while leaving their dominant left paws free to swipe the seal when close enough. That too is based solely on anecdotal evidence.

Sadly, however, there is a great lack of proper scientific investigation of these matters. The paper most often quoted to counter the left-handed polar bear claim is a 2004 study into limb fractures of captive polar bears. In this, the author notes that more injuries occur to the right limbs than the left limbs, from which she draws the tentative conclusion that polar bears use their right limbs more than their left. One might, however, draw the opposite conclusion by suggesting that they are more likely to damage their right limbs because they are more adept at using their left paws. In any case, the number of polar bears in the study was too small to draw strong conclusions.

It is surely time for the scientific and Inuit communities to join forces to resolve this matter. A proper study of handedness in polar bears is long overdue.

223. What proportion of walruses are left-flippered?

In 2003 a team of researchers led by Nette Levermann of the University of Copenhagen, filmed the feeding behaviour of walruses. After dividing the film into 20-second sections, they examined each segment for preferential use of right or left flipper in collecting molluscs and cleaning them before eating. Their results showed that in 89 per cent of the segments in which a flipper is being used, it was the right flipper. This result was widely reported, but with differing interpretations. Some said that all walruses are right-flippered and use that flipper 89 per cent of the time; others said that 89 per cent of walruses are right-flippered, which is much the same ratio as right-handed humans.

To find out which interpretation is correct, I contacted Nette Levermann. She said she didn't know. She was not even sure how many walruses she and her team had filmed, but it was at least five and all were male. Examination of the bones in walrus flippers, however, supports the view that walruses are predominantly right-flippered.

224. What evolutionary advantages arise from right- or left-handedness in humans?

About 90 per cent of humans are right-handed; most of the rest are left-handed, as ambidexterity is rare. Yet why we should have evolved to prefer one hand over the other is difficult to explain. It may have something to do with the use of tools and the acquisition of specialist abilities where there could be advantages in having one hand doing delicate work and the other relegated to basic jobs such as holding or pushing. Or it could have something to do with the development of hemispheric differences in the brain, with certain activities performed by the left-brain and others by the right-brain. Such asymmetric brain development would be expected to lead to differing hand preferences. Yet all such explanations are no more than speculations.

225. Is handedness in some way connected to the acquisition of language skills?

Whatever the reason for having a dominant hand, the next natural question is to ask why most humans favour right over left. Again lateral asymmetry in the brain has been invoked, with the suggestion that right-handedness may be connected with the acquisition of language, as we know that the language centres of the brain are usually on the left side, and the left side of the brain controls the right hand.

In 2011 the psychologist Dr Gillian Forrester published the results of detailed observations of hand usage by a family of gorillas.

While they tended overall to use their two hands equally for tasks she classified as 'social interaction' (such as scratching their head, patting a friend on the back or mothering), they were more likely to use their right hands for inanimate targets (such as using objects and eating or preparing food).

Previous studies on great apes had not identified any consistent population-wide bias for one hand or the other outside humans. The new study, Forrester suggests, may offer evidence that a preference for the right hand in specific tasks may have been the first step in evolving left-brain language skills. 'The basic hierarchy of steps required to make and use tools could be akin to providing us with the scaffolding to build a syntax for language,' she says.

However, while around 10 per cent of people are left-handed, only 5 per cent have their language skills on the right side of their brain. So whether left-handers are right- or left-brained for language skills seems to be equally likely.

See also CATS 72, CHIMPANZEES 77

HUMAN BEHAVIOUR

226. Why do people blush?

We know that blushing is strongly connected to adrenaline. Heightened emotions lead to increased adrenaline levels, which cause the chemical transmitter adenylyl cyclase to send signals to the veins in our face, increasing the blood flow and making us blush. That is a fair answer to the mechanics of blushing, but does not tell us why we blush. What is the evolutionary advantage of having a sympathetic nervous system that tells other people when we are feeling embarrassed? Why is it mainly our faces that blush (though blushing has also been observed to a lesser extent on the neck, chest and even legs)? And why should blind people blush more than sighted people, as some experimental evidence seems to suggest?

227. Why do people cry at emotional times?

Tears come in three basic varieties: there are basal tears, which moisten the eye and protect it; there are reflex tears, which are shed to flush out the eye when it becomes irritated; and most puzzling of all are the emotional tears that flow when we are happy, sad, in pain, or otherwise overcome with strong feelings.

Chemical analysis has shown that emotional tears are higher than the other types in manganese, which is known to be connected with temperament, and prolactin, a hormone that regulates milk production. The natural painkiller leucine enkephalin is also found in tears. Crying may thus help to balance the body's stress levels and provide relief by preventing build-ups of these chemicals, and they may somehow be connected with the reduction of pain. Why this should all happen through tears rather than any of the other, probably more efficient, ways the body has of producing beneficial chemicals and expelling unwanted by-products has not been established.

228. Why do we yawn?

Why we yawn, why other people yawning makes us yawn, and why our yawning makes dogs yawn (→ DOGS 136) are very puzzling questions. One theory is that it has something to do with gasping for oxygen and that it may be a relic from a stage in prenatal development connected to the breathing process. However, experiments in which subjects have been given different amounts of oxygen to breathe have shown no difference in their yawning behaviour, so a need for oxygen seems unlikely to be a cause. Another theory is that it may be caused by a gene that had something to do with the gills of the fish-like creatures from which we are all descended.

One idea that has been supported by experiments is that yawning may help to cool the brain or the body. It has been demonstrated that budgerigars yawn more in a warm room than in a cold one, and that they yawn most of all during periods when the temperature is rising. Human yawning also changes with temperature: people holding a cold pack to their foreheads yawn less than those holding a warm pack. Whether that shows yawning to be a cooling mechanism, however, is arguable.

229. Why do we kiss?

The Austrian psychoanalyst Sigmund Freud (1856–1939) said that kissing has its origins in breast-feeding, specifically the pouting movements that a baby's lips makes when searching for a nipple. Some suggest that it goes back to a time when mothers would chew food for their babies and deliver it mouth-to-mouth.

Another theory is that kissing is part of a tasting and smelling ritual that is useful in selecting a mate. The scent glands on a person's face may help a person select the partner with whom they will have the healthiest offspring. Another idea is that kissing may be a way of demonstrating one's good intentions and earning trust. Just as stroking instead of hitting demonstrates goodwill and sensitivity, kissing instead of biting may give a positive message.

Yet none of this explains why certain cultures, such as Australian Aborigines or the inhabitants of certain South Pacific islands, never kissed until the Europeans arrived and taught them how. Even the Chinese did not habitually indulge in kissing to any great extent until recently. Whether kissing is something in our genes, developed through evolution, or whether it is learned behaviour is still passionately debated.

230. What was the origin of laughter?

Anatomically, laughter is caused by the epiglottis constricting the larynx, but why should our epiglottis constrict our larynx when we are amused, or tickled, or happy, or experiencing a release of tension? Research with bonobos (pygmy chimpanzees from West Africa) shows a strong similarity between the laughter of tickled bonobo infants and the laughter of human babies, and this supports the view that laughter emerged very early in human evolution, before our ancestors diverged from the apes around 4 million years ago. It has recently been suggested that laughter emerged as the communication skills of the brain were evolving. Even before we began to talk, laughter enabled entire groups to share a good feeling, so it may have been one of our earliest social skills. Confirming such a theory, however, is far from easy.

231. Why are we ticklish?

Laughing when tickled is a strange mixture of physiological and psychological responses. Generally speaking, it is the most sensitive areas of the body that are the most likely to be ticklish – but this is not always the case. Feet, for example, are usually more ticklish than hands, even though they are less sensitive. There also seems to be an element of surprise needed for tickling to work properly, which may explain why we cannot tickle ourselves. As with laughter, it has been suggested that tickling is a form of communication, but recent research has suggested that ticklishness may not be confined to humans and apes, and that the vocalizations of tickled rats have a lot in common with those of tickled human babies, which is very difficult to reconcile with any psychological theories of tickling.

232. Why do people like music?

'As neither the enjoyment nor the capacity of producing musical notes are faculties of the least use to man in reference to his daily habits of life, they must be ranked among the most mysterious with which he is endowed.' So wrote Charles Darwin in *The Descent of Man* (1871). In the century and a half that has passed since Darwin made that comment, the mystery of music has scarcely diminished at all.

Music of some sort seems to have existed in every known human culture throughout every age, which leads to the almost inescapable conclusion that an ear for music and a love of music are innate human traits that must have developed at some stage of our evolution long ago. For the 'music gene' to have survived and become so prevalent, it must be beneficial either to survival or to our ability to reproduce. To argue the first case, one would have to show that an appreciation of music somehow made humans better suited to cope with their environment; in the second case, one would need to show that being musical improves a human's chances of finding a mate and/or having healthy offspring.

Some have speculated that the rhythms and varying pitches in music may have a physical effect through setting up sympathetic resonances in the body, but there are no precise suggestions as to how this might work. Darwin thought it so unlikely that music gave any biological advantage that he concluded that it must be some sort of mating display, comparable to birdsong. Others, however, have found that analogy highly unconvincing. The French anthropologist Claude Lévi-Strauss wrote in 1971: 'Although ornithologists and acousticians agree about the musicality of the sounds uttered by birds, the gratuitous and unverifiable hypothesis of the existence of a genetic relation between birdsong and music is hardly worth discussing.' Nevertheless, the idea of music having its origins as a form of sexual display has attracted growing interest in recent years, and when opponents have cited the early death of so many pop stars as an argument against any supposed evolutionary advantage of musicality, supporters of the theory have insisted that the existence of groupies lends weight to its claim to be useful for reproductive success.

In that respect, an analogy between the evolution of human musicality and peacock's tails has been suggested. The peacock's proud display of its tail feathers is a potent display of its strength and masculinity, so females pick the males with the best tails. This not only strengthens the survival value of the good-tail gene, but also favours the peahens who go for males with big tails. The importance of the tail display thus grows, to a point where huge tails are in danger of becoming an encumbrance. In a similar fashion, musical production and musical appreciation feed on one another, leading to ever greater complexity in music and the development of different types of music in different cultures. A tenuous argument perhaps, but probably the best alternative to Darwin's dismissal of musicality as a complete mystery.

233. Why does heat make people aggressive?

For more than a century, a wide range of studies have confirmed that people think and act more aggressively when they are hot. Analysis of violent crime statistics shows higher rates during hot weather than when it is cold, while experiments with subjects in temperature-controlled rooms show that those in the hot rooms have more violent and aggressive thoughts than those who are kept chilled.

Yet other experiments, as well as common experience, indicate that being hot makes people lethargic and lacking energy. This is highly paradoxical: aggression and violence demand energy, so how is it that the same thing can lead to a decrease in energy but an increase in aggression? This sort of question continues to leave students of human behaviour hot and bothered.

234. Does birth order affect personality development?

Writing in 1908, the Austrian psychologist Alfred Adler asserted that birth order played a significant role in determining personality. Firstborn children, he claimed, would have a greater chance of growing up to become leaders, thanks to their being the oldest in the family and having the responsibility of looking after their younger siblings. They would also be more likely to suffer from neuroses and even substance addiction, caused by their feeling of abandonment when the parents started lavishing attention on a younger child.

For similar reasons, Adler said that the middle child in a three-child family would grow up to be the most well balanced and have the greatest chance of success in life, while the youngest would have poor social relationships and be the most likely to take risks, thanks to having been spoiled by his or her parents.

Adler (a second child) did not produce any research to support these assertions, which he felt sure must be true because they fitted with the Freudian model of personality development. A large number of studies have been done over the past hundred years or so to test Adler's claims, but the results have varied greatly, some producing no correlations whatsoever between birth order and personality factors, others producing support for the beliefs, and a few producing results in the opposite direction from those predicted.

Even those who agree that birth order *does* have effects on personality development are unclear to what extent this is due to parental behaviour towards their offspring or whether some of it might be due to an innate predisposition the child is born with.

What wonder grows where knowledge fails.

Tacitus (c. AD 55–c.117)

235. Are humans attracted most to those with a dissimilar genetic makeup?

Choosing a mate can be a problem. On the one hand, it seems best to look for someone like yourself: you will be more likely to empathize with them, less likely to argue, and will, therefore, probably stay together longer. On the other hand, someone just like you is going to have the same weaknesses, making them unable to provide the things that you are least able to do for yourself, leaving the pair of you ineffective in facing the world as a two-person team. Evolutionarily, two sets of dissimilar genes are also more likely to lead to healthy offspring and to maintain genetic diversity in the species as a whole.

Research published in 2009 seems to confirm that for mandrills (baboon-like monkeys, and thus related to ourselves), opposites do indeed attract. The sense of smell is thought to play a large part in mandrill mate-selection, with male mandrills rubbing their chests on trees to release odour from a scent gland. A group of genes called the major histocompatibility complex (MHC) helps to build proteins involved in the body's immune system and also affects body odour by interacting with bacteria on the skin. Female mandrills seem most attracted to males whose smell is least similar to their own, which leads to greater genetic variability in the MHC and therefore produces offspring with a stronger immune system. Female mandrills also tend to mate with as many males as possible, though they show a greater chance of having their eggs fertilized by males who differ from them genetically. If their eggs have a way of rejecting sperm with similar genetic make-up to their own, it would further increase the likelihood of healthy offspring.

We may know more about the mating strategies of mandrills than that of humans, but recent research has shown that men are more attracted to women who are ovulating and that women using contraceptive pills are more likely to favour more effeminate men, while those not on the pill go for butch, muscular types. Since contraceptive pills have been shown to mask a woman's natural smell, it has been suggested that they may be making it more difficult for people to sniff out the right partners.

236. How do we judge whether a picture of a human face is male or female?

One of the things humans are extraordinarily good at is judging whether a picture of a face is that of a man or a woman. Even when the picture is cropped and edited to remove all obvious gender-linked cues such as hair length and make-up, we can usually tell whether it is a man or a woman. For decades, computer programmers have been trying to get machines to do the same thing. A recent paper on the subject mentioned a whole range of techniques and technologies that have been tried, such as principal component analysis, independent component analysis, support vector machines, image intensity. Whatever it is that we do when we see a face, it probably does not include any of the above techniques, but it reaches a decision with great accuracy and speed. Perhaps if we knew how we did it, we would be able to program computers to do it too.

See also GENETICS 208, SLEEP 428–9, 431–2

It is a very sad thing that nowadays there is so little useless information.

Oscar Wilde (1854–1900)

HUMAN EVOLUTION

237. Were the dwarf people of Flores in Indonesia a different species from *Homo sapiens*?

In 2003, on the Indonesian island of Flores, palaeontologists found fossil skeletons of humans who were only 1 metre (3 ft) tall, and who had very small brains. The remains have been dated to between 38,000 and 18,000 years ago, which makes them difficult to fit into the generally accepted picture of human evolution. The genus *Homo* is thought to be between 2.3 and 2.4 million years old. The bipedal *Homo erectus* emerged around 1.5 million years ago, followed by *Homo neanderthalensis* (Neanderthal man) about half a million years ago (→ PALAEONTOLOGY 355–6), while the recent arrival, *Homo sapiens* (modern humans), has been around for only about 200,000 years. Whether Neanderthals and modern humans are both descendants of *Homo erectus*, however, is an open question, and there is still some doubt about whether *Homo neanderthalensis* died out completely, or whether there was some interbreeding with our ancestors – in which case there is still some Neanderthal in our genes.

The fact that the Indonesian 'dwarf' people – known to science as *Homo floresiensis* and to the tabloids as 'hobbits' – are of such recent date poses a real puzzle. The existence of quite sophisticated tools with the fossils suggest that *Homo floresiensis* may have been a sub-species of *Homo sapiens*, yet the considerable difference in height suggests that it was a separate species. There is also the question of how such tools could have been designed by creatures with such small brains. Brain size isn't everything, though. The cranial capacity, and hence brain size, of *Homo neanderthalensis* was as large or even larger than that of *Homo sapiens*. And yet the technologies of the latter, even in its early days, were considerably more sophisticated. Then there is one final question that has not been entirely resolved. While some scientists are adamant that *Homo floresiensis* constitutes a separate species, or at least sub-species, others argue that they are merely specimens of *Homo sapiens* who suffered from a genetic growth disorder.

Interestingly, the island of Flores is also one of the locations where the fossils of dwarf elephants have been found, but the question of a relationship between dwarf elephants and dwarf humans is something beyond even the unknowingness of this book.

238. Why does *Homo sapiens* have forty-six chromosomes while the other great apes have forty-eight?

Somewhere in our evolutionary past, we seem to have lost a pair of chromosomes. Alone among the great apes, humans have forty-six chromosomes arranged in twenty-three pairs, while all the others have forty-eight. Anti-evolutionists have cited this as evidence that apes and humans cannot have had a common ancestor, but in fact the very opposite is true. Chromosome number 2 in humans is almost identical to two ape chromosomes, known as 2p and 2q, spliced together, which strongly suggests that at some stage in our evolution the chromosomes became fused.

Presumably this splicing would have first occurred in a single pre-hominid individual. Under normal circumstances, there would have been a very high likelihood that such a drastic genetic mutation would have died out, but since the fused chromosome of the new model contained all the genetic information of the old, it would still have been able to reproduce with its forty-eight-chromosome neighbours.

The mystery is how the forty-six-chromosome version became a dominant new species. If the fusing together of two chromosomes was not an important part of our transition from ape to human, then one would expect both forty-six-chromosome and forty-eight-chromosome varieties of human to exist today, but if it was a vital change, then what was it about the fusion process that made such a big difference?

239. What is the role of art in human evolution?

Over 17,000 years ago the Cro-Magnons – some of the earliest examples of modern humans in Europe – were responsible for creating a stunning display of cave paintings at Lascaux in France. Many of these paintings are clearly identifiable as animals of the time; some are humans; very few if any are plants or landscapes. The real puzzle, however, is why the Cro-Magnons made these paintings at all.

The usual justification for art of any type is that it is a form of communication; any work of art embodies a crystallization of ideas or emotions or even a broad view of the world into a physical object that others may perceive and share. Yet language is thought to have developed long before the Lascaux paintings, so why invent pictures to communicate when words can do the job already?

The painters at Lascaux were not alone among prehistoric artists. Carvings by Australian Aborigines have been dated to roughly the same period, while a Siberian rock carving of what looks like a man on skis having sex with an elk is thought to date from around 5000 BC.

So what was it all for? Some have suggested that all artistic abilities have evolved as part of sexual selection (→ HUMAN BEHAVIOUR 232); others maintain that art, particularly the ability to draw likenesses of things encountered in the real world, has been an essential component of the evolution of self-awareness and consciousness. One might say it comes down to what you think the Siberian was trying to tell others when he carved that picture in 5000 BC. Was it 'Come and have a good time with me – just look at this if you don't believe me,' or was it 'Bet you can't guess what I was doing this afternoon; look, I'll draw it for you.'

240. Are humans getting cleverer?

According to the New Zealand academic James Flynn, people are getting cleverer. The 'Flynn effect', as it is called, is based on findings in many different parts of the world that people's average scores on intelligence tests have been steadily on the increase. This improvement seems to have been going on since the earliest days of intelligence testing, but whether it reflects an increase in intelligence is highly debatable.

The idea of IQ testing is to measure 'crystalline intelligence', which is the type of pure intelligence supposed to underlie any particular application that requires mental ability. Supporters of Flynn would say that our ancestors, even as recently as our grandparents or parents, had lower crystalline intelligence than we do, though their practical skills may have been just as good. Others, however, would say that all that has happened is that we have become more culturally attuned to intelligence testing, and that IQ tests have become more and more a practical application of an ability that we are all learning at school.

241. Has human evolution stopped?

In 2002 there was an interesting debate held at the Royal Society of Edinburgh under the title 'Is Evolution Over?' The event showed that leading scientists had very different views on the matter, and nothing has happened in the years since that debate to bring us anywhere nearer an answer. There are three main points of view:

(i) Of course evolution is not over. Humans are becoming brighter, healthier and living longer than ever before and evolution is continuing just as it always has.

(ii) Yes, we are living longer, but this is due to increases in our medical knowledge, which is actually working against natural selection. A few generations ago, rates of child mortality were high, which eliminated unfit genes from the gene pool. Now almost everyone (in the Western world at least) survives into adulthood, when they reproduce and their genes, however unfit, live on. Evolution has therefore stopped.

(iii) Actually it's even worse than that. Because of medicine, books, computers and all sorts of other inventions, humans now rely on machines and the ideas of others to give themselves an easy life. A few decades ago, when it was observed that the pigeons in London's Trafalgar Square seemed to be getting weaker and less healthy, part of the blame was put on the tourists who fed the birds, and their tendency to give preferential treatment to injured or unhealthy-looking specimens. Physical unfitness thus came to have positive survival value. The 'evolution is going backwards' school of thought would argue that something similar is happening to the human race.

See also GENETICS 208, 210–2, 214, HAIR 221, HANDEDNESS 224

INSECTS

242. How do fruit flies smell the difference between hydrogen and deuterium?

This may sound rather an esoteric question, but it has important implications for the question of how our sense of smell works. The traditional theory of olfaction is that molecules of a smelly substance dock into receptor proteins in the olfactory membranes, rather like a key into a lock. The docking can only happen if the shape of the odorant matches that of the cavity in the protein; if they fit together, a neural signal is sent to the brain, identifying the smell.

The difference between deuterium and hydrogen, however, is not in the shape of the molecules. Deuterium is an isotope of hydrogen with a proton and a neutron in its atomic nucleus, where hydrogen has only a proton. Their chemical properties are almost identical, and their docking abilities into olfactory receptor proteins ought to be the same. Yet results published in 2011 showed that fruit flies can be trained to distinguish between versions of a fragrant molecule made with hydrogen and those containing deuterium. They can even be trained, by being given mild electric shocks, to avoid either the deuterium-based or the hydrogen-based products.

This result has been seized on by proponents of a new theory of smell that says it is not the shape of a molecule that our receptor proteins respond to, but their rate of vibration. Deuterium atoms, being heavier than hydrogen atoms, vibrate more slowly, but whether that is what the flies are sensing, and whether the human sense of smell works in the same way, are open questions.

243. Why do male crickets that grow up listening to other crickets have better immune systems and bigger testicles than those that don't?

Research at the University of California published in 2010 has shown that the development of crickets is affected by the extent to which they hear other crickets chirping while they are growing up. At first, this was noticed in relation to the size of the cricket community in which they live, but later experiments comparing crickets maturing in a soundproof environment with those that could hear the outside world confirmed that it was hearing the chirping that made all the difference. Specifically, it was discovered that crickets growing up in silence had weaker immune systems than those that grew up hearing chirping. Later, it was also found that the reproductive systems of those in the soundproof crèche were smaller than those in the noisy world.

This does not make bad sense: the extent to which they hear other crickets would under normal circumstances be a good guide to how many other crickets are in the neighbourhood. The more crickets there are around, the more likely a specific cricket is to meet others and catch a disease from them. Also the more other crickets are around, the more chance there is to find others to mate with. So more chirps means more disease and more mating, hence demanding a better immune system and bigger testicles. How auditory signals become translated into such physical changes, however, has yet to be explained. Crickets, incidentally, hear through vibration receptors on the front of their legs.

244. Do flies have free will?

Theologians and philosophers have argued about free will for centuries, but in 2007 scientists reached the conclusion that free will does exist – for fruit flies, at least. The experiment on which this conclusion was based involved a fly glued to a board by its head and with its wings and legs tethered. The tethers were attached to sensitive equipment which could record the strains being applied by the fly to its restraints. The entire apparatus was placed in a draught-free environment so that no external forces were acting on the fly.

The argument was that if the fly was merely a creature that responded automatically to stimuli, the pattern of its tugging should be repetitive or random. Analysis of the tugs of its legs and the flaps of its wings, however, revealed a changing but non-random pattern. In some sense the fly was therefore 'deciding' what to do in a way the experimenters said exhibited free will.

The argument about human free will is certain to go on, but it will now be more difficult to say that humans have free will and flies do not.

See also BEES 34–7, BUTTERFLIES 65–6

INVENTIONS

We know that Adolphe Sax invented the saxophone and that Christopher Cockerell invented the hovercraft, but there are a vast number of inventions, from the abacus to the zip, for which the original inventor's name is lost or credit for the invention is disputed. It would be easy to fill this book with such things, but we must make do with just a handful of examples in honour of all lost and disputed innovators.

245. Who invented the pencil?

Around the middle of the sixteenth century, a graphite mine was discovered high in the hills of Cumbria, England. The substance was not known at the time to be a form of compressed carbon, but it was quickly recognized by local farmers to be very useful for marking sheep. For that purpose, the graphite was cut into sheets from which rods were sliced. Such rods were originally wrapped in string or sheepskin, but later inserted into hollowed-out wooden holders.

Before that time, artists had used similar rods made of lead, which were referred to as 'pencils' from the Latin word for a little tail. (The word 'penicillin', so-called for the tail-like strands of the fungi from which it comes, has the same derivation.) Since graphite was originally thought to be a form of lead, the newly invented tools were called 'lead pencils' – and the name has stuck.

Incidentally, the first person to patent a pencil sharpener was John Lee Love of Massachusetts in 1897.

246. Who invented the screw?

We know that the English instrument-maker Jesse Ramsden invented the first satisfactory screw-cutting lathe in 1770, and that in 1797 Henry Maudslay produced an improved version that allowed mass-production of accurately made screws. However, metal screws and nuts to fit them had been made from the fifteenth century, and wooden screws had been used in wine or olive oil presses for centuries before that.

In ancient Greece, Archimedes had invented a 'screw' for lifting water, but we have no idea who borrowed the screw-shape idea from him to produce a screw that could fasten things together.

247. Who invented the potato peeler?

Even the inventor of the first potato peeler may not have realized that he or she had invented the potato peeler. Most of the early patents for such devices describe themselves as 'apple peelers' or 'vegetable parers', though to modern eyes they may look like potato peelers. The real trouble, though, is that the invention of the potato peeler almost certainly pre-dates the issuing of patents.

The first apple peeler patented in the USA, for example, dates back to Moses Coates in 1803, but no patents were issued before 1790, and instruments with similar functions were almost certainly in use at that time. The *Oxford English Dictionary* gives no citation for uses of the term 'potato peeler' until 1869.

248. Who invented skateboards?

Skateboarding first became a craze in 1963–4 when the first competitions were held and shops experienced huge demand for these platforms on roller-skate wheels. The idea seems to have begun in the 1950s, when Californian surfers latched onto the idea of surfing the streets. Yet nobody knows who started it. Perhaps several people came up with the idea at the same time – certainly several have claimed to be the original inventor, but no convicing evidence has ever been presented.

See also COFFEE 89

JESUS CHRIST

249. What did Jesus do between his childhood years and his late twenties?

Apart from a brief passage in Luke (2:41–52), when the twelve-year-old Jesus goes missing on a trip in Jerusalem and is found in the Temple debating with the elders, there is no reference to anything Jesus did from shortly after his birth until the ministry of his final few years. All Luke tells us is that after he was found in the Temple, Jesus went back to Nazareth with his parents and was submissive to them and increased in wisdom and stature. Considering the dramatic account of his birth, at which he was hailed as the promised Messiah, this is a huge and puzzling gap in the story.

250. Where was Golgotha, the site of Jesus' crucifixion?

According to the New Testament, the crucifixion took place at Golgotha, a name deriving from an Aramaic word meaning 'place of the skull'. Some have suggested this was a hill whose shape was reminiscent of a skull; others suggest it was believed to be the burial place of the skull of Adam. When St Jerome translated the relevant passage into Latin in the fourth century, he used the phrase *Calvariae Locus*, meaning 'place of the skull', and it is from this that the English name 'Calvary' derives. But where was it?

John's Gospel tell us that Jesus' body was carried only a short distance before it was placed in the tomb. This suggests that the site was probably near a cemetery.

Hebrews 13:12 says that the site was 'outside the city gate', but does not specify which gate, while Matthew tells us it was near a road that carried a lot of foot traffic.

The site on which the Church of the Holy Sepulchre was built satisfies all these requirements, but it is only one of several that do so, and Biblical historians are divided in their views as to whether it is in the right place.

251. How tall was Jesus?

Estimates for Jesus' height vary between about 1.38 and 1.85 metres (4 ft 6 in and 6 ft 2 in), which is about as wide a range as one could wish not to have. According to the first-century Jewish-Roman historian Josephus, he was three cubits tall, but there is a good deal of argument about the size of a cubit (→ OLD TESTAMENT 339). The common cubit is generally thought of as 46 cm (18.5 in), which gives us the 1.38 m figure, but there is also the 'royal cubit' measuring about 53 cm (21 in), which would give a height nearer 1.59 m (5 ft 3 in), which was about the average adult male height of the time. On the other hand, measurements based on the image on the Shroud of Turin (→ CHRISTIANITY 80) have led to estimates between 1.75 m and 1.85 m (5 ft 10 in and 6 ft 2 in). There is no mention of Jesus' height in the Gospels, which suggests to some commentators that his height must have been around average.

JUDAISM

252. What happened to the Ark of the Covenant, or was it always mythical?

The idea of the Ark of the Covenant runs through the Jewish Bible with impressive consistency. This casket, in which Moses was said to have been given the Ten Commandments, and which in some versions also contains Aaron's rod and a jar of manna, appears first in the Book of Exodus, and then many times in Deuteronomy, Joshua, Judges, I Samuel, II Samuel, I Kings, I Chronicles, II Chronicles, Psalms and Jeremiah, as well as turning up again in the New Testament in Hebrews and Revelation. All of these attest to the great power of the Ark – not so far different from what Indiana Jones experienced in *Raiders of the Lost Ark*, in fact.

Mere repetition, however, does not make it true, and the Ark has been sought in vain for two millennia, with a multitude of myths locating it in a number of widely separated places in Asia, the Middle East, Europe and Africa. In 2008 Tudor Parfitt published *The Lost Ark of the Covenant*, a book that tells the tale of the author's researches into a clan in Zimbabwe called the Lemba, who claimed to be a lost tribe of Israel. In their myths, their ancestors brought the Ark of the Covenant to Africa from Jerusalem. The claims of the Lemba were generally viewed with intense suspicion until an examination of their genes revealed a marker they shared with a group of Jews.

Following the leads in the Lemba myths, Parfitt discovered an ancient wooden box with certain similarities to the Biblical Ark. Carbon dating revealed this as having been made in the fourteenth century AD, suggesting to Ark-believers that it was a direct copy of the original made when the true Ark destroyed itself.

253. How did the Jewish people split into Ashkenazim and Sephardim?

Following the conquest of the kingdom of Israel by the Assyrian empire around 720 BC, many Jewish people dispersed throughout the Middle East and North Africa. The Jewish-Roman wars and persecution by various Roman emperors led to the exile of the remaining Jews from Judaea in the second century AD and further dispersion. This diaspora led to the formation of small, initially isolated Jewish communities throughout the known world, which eventually coalesced into two main groups: the Ashkenazim, essentially comprising the European Jews, and the Sephardim, who were the Asian Jews.

That, at least, is the traditional explanation of both religious and cultural differences between the groups, as well as their general appearance. Yet recent genetic studies have produced a number of surprises. While confirming that both groups had the same origin, analysis of the DNA of both Ashkenazi and Sephardic Jews has revealed an unexpected degree of similarity. Many Jewish communities were thought to have been founded by converts to the faith; that view is not supported by genetic analysis. In particular, the role of the Khazars (→ ANTHROPOLOGY 28) in the development of Judaism in Eastern Europe is left in great doubt by genetic evidence.

As more and more studies of the gene pools of various Jewish communities are completed, a growing picture is emerging of their relationship to one another and the extent of intermarriage between existing communities. In some cases, this confirms historical views, in other cases it seems to contradict them. The jigsaw of the Jewish diaspora is far from complete.

254. What happened to the Lost Tribes of Israel?

Now here is a Biblical mystery that modern science really is beginning to throw some light on. According to the Book of Genesis, the twelve sons of Jacob fathered the twelve tribes of Israel, which settled on the two sides of the Jordan River. After the death of King Solomon, the tribes of Judah and Benjamin formed the Southern Kingdom of Israel, the other ten tribes the Northern Kingdom. In 722 BC the Assyrians conquered the Northern Kingdom and much of the Southern, and sent the ten tribes into exile.

That, in a contentious nutshell, is the story behind the Ten Lost Tribes of Israel. Some accounts speak of a total of thirteen tribes, as Joseph's two sons are said to have split into two tribes, while others hold that the tribe of Levi, whose members were to be the priests, had a different status from the others. But however one counts them, there has long been a mystery about what happened to the Ten Lost Tribes.

Genetic studies of Jewish populations in various parts of the world are beginning to piece together the picture. The priestly origin of the tribe of Levi seems to be confirmed by the existence of a so-called 'Cohanim' gene, which has been found in members of the priestly class (and those named Cohen) in many different Jewish communities. Scientific studies have shown that the Lemba tribe of Zimbabwe, who claim to be descended from one of the Lost Tribes (→ JUDAISM 252), began to diverge genetically from Middle-Eastern Jews around the time of the Assyrian conquest of Israel.

The first pieces of the genetic jigsaw are in place. Whether the final picture will reveal twelve tribes, which date back to around three thousand years ago as the Biblical story suggests, is an intriguing question.

See also OLD TESTAMENT 339–43

LANGUAGE

255. At what point did early humans begin to use language?

Somewhere in the long history of human evolution, we began to talk to each other, but estimates of when that point came vary vastly. Some say that it began with the emergence of *Homo erectus* around 2 million years ago; others say that the shape of the ears of *Homo heidelbergensis*, dated to about half a million years ago, show that he was the first to speak; while the most conservative view is that proper language did not come into being until *Homo sapiens sapiens*, the truly modern human, came into existence 200,000 years ago.

The main reason for the great disparity is the underlying question of what we mean by language. Vocal communication is generally taken as the start of language (as written communication seems to demand conceptual skills of a higher order), but a high level of structure is needed before such behaviour can be called 'language'. Birds and animals, for example, have been shown to have vocabularies consisting of a number of 'words' or grunts with different meanings, but this is hardly language.

Whether such squawks and grunts developed seamlessly into language as we now know it, or whether an evolutionary change was responsible for the introduction of syntax and the level of abstraction that exists in all human languages, is an open question.

256. Did Neanderthals talk to each other?

Until the 1980s the popular image of Neanderthal man was that of a great, unkempt, low-foreheaded, lumbering creature who uttered nothing but grunts. In 1983 that view changed with the discovery of a Neanderthal hyoid bone in a cave in Israel. The hyoid is a small bone that connects the muscles of the tongue and the larynx, thus giving the possibility of the articulation of a wide range of sounds that had previously been impossible. Lumbering, unkempt and low-foreheaded he may have been, but Neanderthal man had the physical means of talking much as modern humans do today. Since analysis of the ears of his predecessor, *Homo heidelbergensis*, had already shown that early humans had the acoustic apparatus needed to distinguish a wide range of sounds, Neanderthal man began to look like a good candidate for the first talker.

In his 2006 book *The Singing Neanderthals*, the palaeontologist Steve Mithen has developed this theory to suggest that language developed from song, and that the Neanderthals were the first to make that transition. He even coined a word for their humming language, which he called 'hmmmmm' because it would have been 'holistic, manipulative, multi-modal, musical and mimetic'. 'Hmmmmm' could also be taken as the reaction expressed by others in the field to this intriguing theory.

257. Who were the speakers of Proto-Indo-European?

In the late eighteenth century the English scholar and philologist Sir William Jones commented on the similarity between certain words in Sanskrit, one of the ancient languages of India, and their equivalents in classical Latin and Greek. Ever since then, interest has focused on the supposed common root of all the languages of the so-called Indo-European family, which includes not only most of the European languages, from Gaelic and Greek to Spanish and Swedish, but also many languages in India and Iran. Around half the world's population are now thought to be native speakers of languages that developed from a Proto-Indo-European root, though no records of this root language survive. Indeed, it is thought to have been purely a spoken language, never written, but the question of who spoke it is also unanswered.

By 3000 BC Proto-Indo-European (PIE) had begun to diverge into other languages, but what happened before that, and where PIE came from in the first place, is a matter of pure speculation. Precise suggestions for the time of its origin range between 4000 BC and 7000 BC, but some have suggested that the birth of PIE was several millennia earlier than that. Where did it start? Armenia, Eastern Europe, India, Northern Europe and Anatolia have all been suggested, but it is little more than educated guesswork. All we do know, from the words and sounds shared by many languages that developed from PIE, is that they probably used solid wheels without spokes, travelled in boats, had snow in winter and worshipped a sun god.

258. Where did the Basque language come from?

Basque is a real linguistic mystery. Spoken by around 650,000 people in northern Spain and southwestern France, it appears to be related to no other language. The Basque region is surrounded by an area dominated by Romance and other Indo-European languages, to which it bears no similarity. The general conclusion is that Basque is the last survivor of a language group that pre-dated Indo-European, and that somehow the Basques were ignored by the Romans as their military and linguistic conquests spread through Europe.

Another theory is that the Basque people and their language came to the region from somewhere further east, after the fall of the Romans. Having nothing to compare it with, however, the history of the Basque language has been impossible to trace with any degree of confidence.

259. Is an infant's ability to learn language in some way pre-programmed into his or her brain?

From the 1950s onwards, the American linguist-philosopher Noam Chomsky developed a theory that language skills are innate in humans; in other words, that we are genetically pre-programmed with a capacity for learning languages. In Chomsky's model we are all born with a 'language organ' in the brain that has evolved for the specific task of learning and using language. While we acquire a specific language early in life from our parents or those around us, we are all equipped with the same mental skills for what he called 'universal grammar', which includes the concepts of syntax and the principles of communication skills that underlie all languages.

Chomsky supported his theories with evidence from the formation of human speech organs, the speed with which children acquire their mother tongue, the arguably unique language skills of humans, and studies of the mistakes made by children in their early efforts at speech. When a child says something in a logical but ungrammatical way – such as arguing with anyone who says 'Chomsky wrong' by saying 'No Chomsky wrong' instead of 'Chomsky not wrong' – Chomsky would, after patting the child on the head, no doubt commend her use of universal grammar while waiting to acquire the specific syntactical rules of English.

Attempts to isolate this 'language organ' or to draw up the rules of universal grammar, however, have not made much progress. Opponents of Chomsky's innatist theory would deny that language is so special. Our brains have an undeniable ability to recognize patterns and form concepts, and that, they would contend, is enough to explain language acquisition.

260. To what extent does language influence thought?

One of the most intriguing and contentious debates in twentieth-century linguistics centred on the Sapir-Whorf hypothesis. This was something of a misnomer, as the anthropologist Edward Sapir had little to do with it, while Benjamin Lee Whorf was a Massachusetts fire inspector who only dabbled in linguistics as an amateur. What's more, the whole thing wasn't really a hypothesis in the first place.

Now known by the more academically respectable title of 'the principle of linguistic relativity', the idea comes in two varieties. In the strong version, language totally determines thought, and linguistic categories limit and determine cognitive categories. In the weak version, language and linguistic categories influence thought and some kinds of non-linguistic behaviour.

Whorf gave these ideas great popularity in an article in 1940 in which he mentioned the number of words Eskimos allegedly had for snow. Looking for an example of language influencing perception, he listed seven different types of snow for which he maintained there were different Inuit words. It is highly doubtful, however, that he ever knew a single word in any Eskimo dialect, or that he had ever met an Eskimo. Nevertheless, the idea caught the public imagination, and, thanks to this article, Eskimo words for snow rapidly became the standard example of the Sapir-Whorf hypothesis, with the number of such words apparently increasing every time it was mentioned.

The Scottish-born US linguist Geoffrey Pullum gave a hilarious account of the Eskimo snow-word inflation in *The Great Eskimo Vocabulary Hoax and Other Irreverent Essays on the Study of Language* (1991). He ends the title essay with an account of his own actions on hearing a lecturer perpetrate the 'Eskimological falsehood', and tells the reader what to do if it should happen to them:

> *I just held my face in my hands for a minute, then quietly closed my binder and crept out of the room.*

> *Don't be a coward like me. Stand up and tell the speaker this:*
> *C.W. Schultz-Lorentzen's Dictionary of the West Greenlandic*
> *Eskimo Language (1927) gives just two possibly relevant roots:*
> *qanik, meaning 'snow in the air' or 'snowflake', and aput,*
> *meaning 'snow on the ground'. Then add that you would be*
> *interested to know if the speaker can cite any more.*

For the record, the strong principle of linguistic relativity has been generally abandoned after research has shown that people's perception of colour differences appears not to be influenced by the words for colours in their language. But the extent to which the weak principle applies is still a matter of debate.

See also ENGLISH LANGUAGE 176–81, WRITING SYSTEMS 493–5

MAGNETISM

261. What is the cause of the Earth's magnetic field?

Ever since the Chinese watched the behaviour of slivers of lodestone in bowls of water over a thousand years ago, we have been aware of the Earth's magnetism. Around 1600 the English physicist William Gilbert showed that this magnetism comes from the Earth itself, and some two hundred years later, the German scientist Carl Friedrich Gauss showed that the source of this magnetism is right at the centre of our planet. Quite what causes it, however, has remained a challenging problem.

In 1939 the US physicist Walter Elsasser proposed the idea of a dynamo mechanism caused by convection currents in the molten iron of the Earth's core combined with the Earth's own rotation, but his theories never quite tallied with our knowledge of convection and our growing understanding of the composition of the Earth's core. In 2007 another US scientist, J. Marvin Herndon, proposed that a uranium-based natural nuclear reactor at the Earth's centre could be a possible source for the magnetic field, and the dynamo and geo-reactor theories have been fighting it out ever since.

Whichever theory wins the argument, it will also have to answer this question:

262. Why does the Earth's magnetic field sometimes flip, changing North and South Poles around?

Every million years or so, the North and South Magnetic Poles switch, and the Earth's magnetic field is reversed. That is the astounding conclusion drawn from the analysis of igneous rocks. Such rocks, formed from the molten state, contain indicators of the magnetic field at the time of their solidification. These magnetic 'fossils' indicate that reversals of the Earth's magnetic poles have occurred, but no one has yet established how this change occurs, or how long it takes for the change to be accomplished.

263. What effect would a flipping of the planet's magnetic poles have on life on Earth?

The short answer is that we haven't the faintest idea. We know the Earth's magnetic field plays a part in protecting us from solar radiation, and there is no reason to believe that would change if the poles were reversed, but what would happen during the period of reversal, which some suggest could last for 5,000 years, is completely unknown.

There is also the matter of those birds and animals that make use of the magnetic field for purposes of navigation. Whether a slow change in the field would fatally confuse such creatures, and whether the time frame of magnetic flipping would allow an evolutionary change to take place that would provide the necessary adjustment, are interesting questions.

As for human beings, since we no longer use the Earth's magnetism to steer ships, it is quite likely that we would not even notice. That is, if any of us are still around when it next happens.

See also BIRDS 47

MAMMALS

264. What is the relationship between the size of a mammal and the amount of food it requires?

Pound for pound, large animals need less food than small animals. It has been said that this is because small animals run around more, and therefore need more energy. Since the reason given for them running around is to search for food, this looks like a rather circular argument. A more sensible suggestion was made by the German biologist Max Rubner in 1883, who suggested that a large proportion of a mammal's energy requirement came from its need to maintain a constant body temperature. Since heat is constantly being lost through the skin, he proposed that the minimum energy requirement should be proportional to the mammal's surface area.

As mass (m) is proportional to volume, which is proportional to the cube of linear dimensions, while surface area is proportional to the square of linear dimensions, Rubner proposed that the basal metabolic rate (BMR) ought to be proportional to $m^{2/3}$.

That calculation, however, was mainly theoretical, and work in the 1930s and 1940s produced figures based on experiments with living animals suggesting that 2/3 was too small an exponent in the equation and that BMR is proportional to $m^{3/4}$, a figure that appeared to tally well with a famous 'mouse-to-elephant curve' produced by the biologist Samuel Brody at the University of Missouri in 1945, giving the BMR of a wide range of mammals.

Subsequently, the 3/4 figure was generally accepted until a paper in 2003 cast doubt on the earlier research, suggesting that it was weighted heavily in favour of domestic pets and laboratory animals, which would be expected to be less active than specimens in the wild. After adjustment for these factors, the figures seemed to support the earlier 2/3 figure again, so we would have to say that the jury is still out.

265. How far can rats count?

This matter has been hotly debated since the 1970s, and a series of experiments have been designed to test rats' counting abilities. An early experiment had rats running round a course and being rewarded when they did so, except on every fourth run. Results showed that the rats were slower on the non-rewarded runs, apparently showing that they could count up to three. But were the rats really counting, or going by the amount of time it took to run the course three times?

In 1983 H. Davis and J. Memmott performed an experiment giving a fixed number of electric shocks to rats over a variable time period, to see if they could count the number and then realize they were safe. Their conclusion was that 'rats may be taught to count, but such behaviour is highly unnatural and may be blocked or overshadowed by more salient sources of information'. Later the same year, however, the Japanese scientists H. Imada, H. Shuku and M. Noriya performed a similar experiment and concluded that 'there was no evidence that rats could count'.

The general opinion now seems to be that rats *can* count, but that they are extremely reluctant, and will do so only if all else fails. Getting them to count numbers greater than five may be more than they are willing to tolerate. Ants, on the other hand, seem to be very good at counting and can even do simple sums. Their ability to count their walking paces was demonstrated in an experiment on Sahara ants in 2006 in which some ants were fitted with stilts and others had half their legs amputated. Results of experiments published in 2011, meanwhile, revealed ants' arithmetic ability by showing how they can assess optimal routes to find food in a maze.

266. Why do zebras have stripes?

Charles Darwin and Alfred Russel Wallace, nineteenth-century contemporaries and co-founders of the theory of natural selection, argued about this. Wallace said that zebras' stripes made them harder for predators to detect at dusk, when the zebras went to drink. Darwin said that was nonsense, and that the stripes would have provided no protection at all – but he did say he thought they were beautiful.

Other suggestions are that the stripes act as camouflage in long foliage; that stripes confuse predators, who find it difficult to identify the vulnerable young zebra in a mass of stripes; that stripes are a sort of barcode enabling zebras to recognize one another; that black-and-white stripes in general act as a warning signal; that stripes make it difficult for predators to tell where the animal's outline ends; or that black stripes and white stripes absorb different amounts of solar radiation, so setting up a cooling convection system on the animal's skin.

See also PANDAS 357

267. Why do reindeer eat hallucinogenic mushrooms?

There are numerous accounts of animals, from shrews to elephants, gorging themselves on fermenting fruit and displaying symptoms of drunkenness as a result, but there has been little evidence that such animals deliberately get drunk. In the Christmas 2010 edition of the *Pharmaceutical Journal* the deputy editor, Andrew Haynes, wrote of the passion reindeers show for fly agaric – the *Amanita muscaria* mushroom – which is renowned both for its toxicity and for its psychedelic properties. Reindeer have long been known to hunt specifically for this type of mushroom, and Haynes suggests that they deliberately consume a mind-altering substance in order to escape the tedium of the long winter nights. That would imply that reindeer have a level of consciousness and self-awareness rather higher than we generally give them credit for. It might also, as Haynes suggests, give credence to the view that Santa's reindeer really could fly.

268. What do sheep see in other sheep than enables them to recognize each other?

Research has shown that sheep can not only distinguish between other sheep from pictures of their faces, but can remember the faces of fifty other sheep for up to two years. They can also recognize a picture in profile of another sheep's face when they have previously only seen it from the front. Given the choice between a picture of a sheep that is experiencing stress and an unstressed sheep, they will also tend to choose the unstressed one. Mother ewes, however, have difficulty recognizing their own lambs at a distance when the appearance of the head region is altered with the use of coloured dyes.

None of which tells us what is going through the sheep's mind when it looks at these pictures, or what characteristics in a sheep's face it uses in order to make such discriminations.

269. Do monkeys change their behaviour when they know they are being watched?

We are all liable to act differently if we know we are being watched, but do monkeys change their behaviour if they know humans are watching them? The question is important for research into animal behaviour, and a partial answer was given in a paper in 2010. By comparing the movements of a group of capuchin monkeys that knew experimenters were watching them with the movements of a group that were being tracked by radios in their collars, researchers were able to show that 'Capuchins did not move faster, stop to rest less frequently, or display higher levels of activity when they were being followed compared to when they were alone.' They may, however, have changed their behaviour in more subtle ways, or be behaving atypically because they were wearing radio-collars, or because they were already habituated to the experimenters.

See also AARDVARKS 1, ARMADILLOS 29, BATS 33, CATS 72–3, CHIMPANZEES 77–8, DOGS 136–8, GIRAFFES 215–16, PANDAS 357–8, SLEEP 430, SQUIRRELS 453

MARINE LIFE

270. Do fish feel pain?

The question of whether animals in general feel pain dates back to the seventeenth-century French philosopher René Descartes, who was firmly of the opinion that they did not. Animals, he said, lack consciousness so cannot be aware of pain. As our understanding of fish physiology has grown, however, it has become increasingly clear that a fish's response to pain may not be so different from that of a human. When fish are subjected to physically damaging experiences, their bodies respond in ways analogous to those of humans, as does their behaviour. In the late 1990s it was even discovered that fish possess nociceptors in their skin, the pain-detectors that transmit messages to the brain when injury occurs.

So the fish's body responds to the pain, and the fish brain knows about it, but whether the fish itself knows it is in pain may be another matter. Pro-fishing scientists argue that it is the highly developed neocortex of our brain that is responsible for our being conscious of a feeling of pain and other conscious emotions and sensations. The fish brain does not have a neocortex, so if it does experience pain, it must do so in a way that is different from humans. So to answer the question, we would have to know what it is like to be a fish, which may be even more difficult to answer than what it's like to be a bat (→ BATS 33).

271. Where does the nautilus lay its eggs in the wild?

There is hardly any creature known to science that is more mysterious than the nautilus, a relative of the octopus, possessed of numerous tentacles and a coiled shell. These ancient sea creatures, according to the fossil record, have hardly changed in the last 500 million years. They live at depths of around 300 metres (1000 ft), coming up to about 100 m (300 ft) to feed, which they only have to do about once a month. The nautilus moves by a sort of jet propulsion, sucking in water and expelling it at high pressure. Almost blind, they locate food and potential mates by smell or other chemical cues. Some nautilus eggs have been found attached to rocks at their feeding depths, but whether that is their normal egg-laying behaviour is unknown. Like so much else in the depths of our oceans, the nautilus is a mystery. We do not know how long they live; we do not know how long they take to reach maturity after hatching; no single nautilus in the wild has ever been tracked from birth to death. We do not even have any idea of how many nautiluses there are in the oceans, or whether they ought to be classified as endangered. Knowing where they lay their eggs would be a very useful place to start for increasing our knowledge of these elusive creatures.

272. Where in the sea do giant squid live?

The nautilus, however, only gets the silver medal for aquatic mystery. In terms of human ignorance, it comes in some way behind the giant squid. In the ancient world, both Aristotle and Pliny the Elder wrote about these creatures, but the scientific community did not fully accept their existence until the late eighteenth century. Since then, around six hundred specimens have been found, almost all dead. Many had been washed ashore, some were floating in the ocean, others were found in the stomachs of sperm whales.

The first time a live specimen was filmed was in 2001, and the first to be photographed in deep water was in 2004. On the rare occasions when a live giant squid has been captured, it has not survived.

We do not know how many species of giant squid there are: estimates vary between one and twenty. We do not know how they mate. We do not know how big they can grow; corpses and captured specimens suggest that a body length of 5 metres (16 ft) is rarely exceeded, though alleged and probably exaggerated sightings have claimed lengths up to 20 metres (65 ft), including tentacles. We do not know how long they live. We do not know where they live, but they have been found in the Atlantic, Pacific and Southern Oceans.

The even bigger, and just as mysterious, colossal squid has only been sighted off Antarctica. With fewer than ten specimens ever having been found alive or dead, estimates of the size of the colossal squid have been based mainly on tentacles and beaks found in the stomachs of whales, but a recent colossal squid corpse was found with eyes 27 cm (11 in) across, which are the largest eyes ever seen in an animal. And they may well have been larger still when the animal was alive. A maximum length (including tentacles) of 12–14 metres (40–45 ft) is generally thought likely.

273. Why do narwhals have spiral tusks?

Imagine a cross between a unicorn and a whale and you have the narwhal. Or a male narwhal at any rate, for only the males possess the long, straight tusk that extends from its upper left jaw in a left-handed spiral. Females may have shorter tusks. In rare cases, narwhals develop two tusks, but when they do, they both spiral in the same direction.

Darwin was fascinated by the narwhal's tusk, as it seemed to serve no useful purpose. Narwhals have very rarely been seen using it in a fight, or to break ice, or as a tool for feeding purposes. The only suggestion that Darwin and others have made is that it is a secondary sexual characteristic. Whether female narwhals go for a male with a long tusk has yet to be confirmed, but males have been observed rubbing their tusks together in an activity called 'tusking', which is thought to be part of a ritual establishing and maintaining social dominance.

274. Why do whales strand themselves on beaches?

Why single whales, or entire pods numbering in their dozens, strand themselves on beaches and die is one of the great puzzles of the animal world. This happens to around two thousand whales every year, and although a variety of explanations have been suggested, nothing is known for sure. Here are a few of the suggestions:

(i) It all starts when a single sick or injured whale swims into shallow waters. It is then followed by members of its pod, who refuse to abandon it. They are all then trapped by falling tides.

(ii) Their guidance systems are disrupted by naval sonar, or by weather conditions, or by variations of the Earth's magnetic field, or by disease.

(iii) Undersea seismic activity creates currents that lead the whales astray.

In 2000 the US Navy did admit responsibility for intense underwater sonar waves that may have led to the stranding of whales of several species in the Bahamas, but many such strandings are not linked to any known sonar activity.

275. Are jellyfish colour-blind?

Box jellyfish are very active and agile swimmers, and are good at avoiding obstacles on the seabed. This isn't perhaps so surprising as they do have twenty-four eyes. These come in four types, and research in 2007 revealed that it was the lower lens eyes, one of the four types, that are responsible for obstacle avoidance.

The jellyfish detect plants and other potential obstacles by the visual contrast between the object and its surroundings. The question was whether the visual contrast involves intensity or colour. As it turned out, the strength of the jellyfish's response to an obstacle was found to depend only on intensity contrast. According to the researchers this 'fits with our other data which strongly suggest that the jellyfish are, in fact, colour-blind'.

276. How intelligent are dolphins?

The belief that dolphins are highly intelligent all began with measurements of brain size. Large animals, of course, tend to have larger brains than small animals, so the view was taken that what mattered was the 'encephalization quotient' (EQ), which is the ratio of actual brain size to predicted brain size for an animal of the same mass. Cats have an EQ of 1, rabbits and rats 0.4, and chimpanzees between 2.2 and 2.5. At the top of the scale come humans, with an EQ of around 7.5, while bottlenose dolphins come second at 5.6.

This seemed to confirm that dolphins are pretty smart, but in 2006 Paul Manger, a researcher at Witwatersrand University in South Africa, came up with an alternative hypothesis: the dolphin does not use most of its brain for thinking, but rather for thermoregulation. Dolphins are, after all, warm-blooded creatures living in cold waters, which get even colder when they dive. Maintaining their body temperature is vital and the dolphin brain might not be so much a thinking organ as a glorified thermostat.

At the start of 2010, scientists in the USA declared that dolphins should be treated as 'non-human persons' in view of their intelligence. Their argument, based on brain size, is less convincing than behavioural evidence would have been. Dolphins have, after all, been observed to do some very clever things, such as recognizing themselves in mirrors, holding sponges over their noses to protect themselves from spiny fish, and teaching tricks learned from humans to other dolphins – all types of behaviour once thought to be exclusive to the great apes.

277. Can lobsters recognize other lobsters by sight?

For some years, it has been known that lobsters can recognize each other by smell. Lobsters tend to urinate most when they are aggressive, and when two lobsters fight, both remember the smell of the other's urine. If the same lobsters encounter each other again, the one that lost the fight will avoid any altercation with the lobster that beat him, while the winning lobster will be eager to fight again. Since lobsters have poor vision and fights tend to take place in the murky depths of the sea, it is hardly surprising that smell rather than vision is the dominant sense. More recently, however, the question has been raised as to whether lobsters know each other's faces.

In order to answer this question, researchers at the University of Florence, Italy, placed ninety-eight lobsters bought in a local fish market in a tank divided by a variety of partitions. These partitions could be opaque, transparent, perforated (to allow the transmission of smells) or unperforated. The results showed that when lobsters could see each other, they started bumping against the divider aggressively, but when they could not see each other, they hardly moved at all, whether or not they could smell the others. When the partitions were removed, the lobsters that had previously seen each other started fighting or actively avoiding each other, while the others adopted a more investigatory approach.

The conclusion was that lobsters can recognize by sight when another lobster is around, but whether they can recognize a particular lobster they have met before by sight alone is still an open question.

See also THE PLANETS 377

MATHEMATICS

278. Are there any odd perfect numbers?

A perfect number is defined as an integer that is equal to the sum of its proper divisors. The first perfect number is 6, as the divisors of 6 are 1, 2 and 3, and 6 = 1+2+3. The next perfect number is 28 (as 1+2+4+7+14 = 28), and the next two after that are 496 and 8,128. All these were known to the ancient Greeks, but it was not until 1456 that a fifth perfect number, 33,550,336, was discovered.

In the eighteenth century the Swiss mathematician Leonhard Euler proved that every even perfect number has the form $2^{p-1}(2^p-1)$, where p is a prime number. The five examples above are given by the values 2, 3, 5, 7 and 13 for p. It is not known, however, whether there are infinitely many perfect numbers, or whether there are any odd perfect numbers. The best we can say is that if there is an odd perfect number, it must have at least 1,350 digits.

279. Are pi and e absolutely normal numbers?

To explain what this is about, let's start with the first thirty digits in the value of pi (π):

π = 3.141592653589793238462643383327...

Now it has long been known that this decimal expansion goes on forever and does not endlessly repeat the same string of digits. But the question of whether it is 'normal' or 'absolutely normal' depends on the relative number of occurrences of each digit in the expansion. So let's count how many times each of the digits from 0 to 9 occurs. There are no zeroes, two 1s, four 2s, seven 3s, three 4s, three 5s, three 6s, two 7s, three 8s, and three 9s, but if we went on and extended the value of pi to a hundred or a thousand or a million places or more, we would find that the numbers of each digit become closer to each other. Not only that, but if we counted the number of times the sequences '11' or '90' or '47' or any other two-digit number appears, they too would be roughly equal. And the same would be true of any other number of digits.

Any infinite decimal that has the property that any string of digits is as likely to occur as any other string of the same length is called a 'normal' number.

The above example, however, was linked to the decimal expansion for pi; that is, the one based on our usual counting system to base ten. What about the binary pi:

π = 11.00100100001111110110101010100010001000001010101...?

Does that have the same number of 1s as 0s, and the same number of 00s as 01s or 10s or 11s? And what if we had used base three, or four, or some other number?

Any number that displays the quality of being normal in any

system we choose is called 'absolutely normal'. The funny thing is that although mathematicians have proved that almost all numbers (in a very precisely defined sense of 'almost all') are absolutely normal, we can hardly give an example of any of them. Nor is there any known method of determining whether a given number, such as pi or *e*, the base of natural logarithms, is absolutely normal.

280. What is the answer to the Collatz problem (or Syracuse algorithm)?

Think of a number. If it is even, halve it; if it is odd, multiply it by three and add one. Then apply the same process to the result and keep on doing so. So if we start with 11, for example, the sequence goes: 11 - 34 - 17 - 52 - 26 - 13 - 40 - 20 - 10 - 5 - 16 - 8 - 4 - 2 - 1 - 4 - 2 - 1... and then keeps repeating 4 - 2 - 1. The conjecture is that whatever number you start with, it always ends up at this 4 - 2 - 1 sequence. Every number so far checked has eventually reached that end, but there is no known proof that all numbers do so.

Warning: It is known that if there is any number that does not eventually fall into the 4 - 2 - 1 pattern, then it must have at least nineteen digits, so don't waste your time looking for a smaller one.

281. Does P = NP?

In 2000 the Clay Mathematics Institute in Cambridge, Massachusetts, issued a list of seven Millennium Prize Problems comprising what they considered to be the most important unsolved questions in mathematics. A correct solution to any of them will win a prize of $1 million. So far, only one has been solved, and the mathematician responsible turned down the prize. The P = NP conjecture is one of the six remaining problems. Being at a stratospheric level of mathematics, none of the Clay problems are easy to explain, but here is an idea of what P = NP is all about.

It is all to do with the time it takes a computer to find the answer to a question. Suppose we wanted to know what is the longest word in this book. All a computer would have to do is look at each word in turn, count how many letters it has, and see whether it is longer than the longest one it has previously encountered. The number of operations the computer has to execute would be roughly proportional to the number of words in the book, which we shall call N. Other more difficult problems might require a number of operations proportional to N^2 or N^3 or some other power of N. Those would be much bigger numbers than N, but all such powers soon become insignificant, as N grows, when compared with 2^N.

If a problem can be solved with a number of operations given by powers of N, it is said to be solvable in polynomial time; if it needs 2^N operations, it takes exponential time. The implications for computing can be very important. If N = 100, for example, and our computer can perform 100 operations a second, then it will take just over a minute and a half to complete N^2, and almost three hours for N^3, but 2^N would take it longer than the universe has been in existence.

So back to the P = NP problem, in which the right-hand represents all the problems whose answers can be found in polynomial time, while the left-hand side is all the problems whose answers could be checked in polynomial time. There is a big difference between finding and checking an answer. Finding the factors of a large number can be a very tedious and time-consuming process. Checking that two particular numbers are the factors you seek is easy: you just multiply them together.

There are many such problems with easy-to-check answers, but where only exponential-time computational methods are known for finding those answers in the first place. If P = NP is correct, then there must be a polynomial-time method that nobody has thought of yet. And in some cases, that would be worth a million dollars of anyone's money.

See also GAMES 203, NUMBERS 337, PRIME NUMBERS 390–2

MEDICINE

282. Does the placebo effect exist, and if so what causes it?

Ever since the eighteenth century, the medical effect of placebos has been remarked upon. In recent times, acceptance of the 'placebo effect' has been sufficiently widespread to ensure that any test of a new drug will include a group of subjects given, in place of the drug being tested, a placebo – something that looks just like the genuine drug but has no physiological effect. More often than not, the placebo will be found to have had a beneficial effect on a number of patients. In the original clinical trials of Viagra as a treatment for erectile dysfunction, for example, around 25 per cent of the subjects who had been given a placebo reported significant improvement (compared with over 70 per cent for those who took the real drug, which accounts for its immediate popularity).

In recent years, however, several studies have cast doubt on the placebo effect, with suggestions that its potency may be limited to ailments with a psychological or subjective side to them, such as those involving the patient's perception of pain. When a physiological effect is associated with taking a placebo, it may be connected to the production in the brain of dopamine, which is known to be associated with a feeling of well-being. Curiously, the medical profession has always been hostile to the idea of faith healing, yet the effect of faith to a believer ought to be very similar to that of a placebo on someone who thinks he is being given medicine.

The whole picture, however, has been thrown into additional confusion by a recent report of a trial in which patients suffering from irritable bowel syndrome were given placebos and told that they contained no drugs of any use whatsoever. And they still reported improvement in their condition. So placebos seem to work even if you know they are placebos. Whether belief in the placebo effect is necessary for them to work, however, requires further research.

283. Is there any scientific basis for homeopathy?

Can a solution of some ingredient have beneficial effects even if that solution has been diluted to a degree at which not a single molecule of the ingredient is still present? That idea, enshrined in homeopathic theory and practice, has been roundly condemned as preposterous by mainstream medical practitioners and scientists, while studies have repeatedly reported that homeopathic treatments achieve no better results that placebos. Yet the homeopathic industry continues to thrive, and both practitioners and many of their patients insist that it works.

From a strict scientific point of view, the case for homeopathy must be considered unproven, yet in 2010 a Nobel prize-winning scientist announced a finding that was seized upon by the homeopathic brigade as supporting their claims. Luc Montagnier, the French virologist who won the 2008 Nobel prize for physiology or medicine for linking the HIV virus to AIDS, reported a series of experiments investigating the electromagnetic properties of highly diluted biological samples. He found that small DNA fragments caused identifiable electromagnetic signals in aqueous solutions in which they were dissolved, and those signals continued to be detected even after the solution had been diluted to a very high degree. He suggested that subatomic structures were created in the water that persisted even when the substance that created them was removed.

Professor Montagnier did not mention homeopathy in his report, but supporters of alternative medicine claimed that it vindicated all they had been saying about water having a memory, and that

Montagnier had now explained just how homeopathy works.

Or it might just be the placebo effect.

284. Why does honey help wounds heal?

As long ago as two thousand years before the discovery of bacteria, honey was used to treat wounds infected by them. The Greek physician Pedanius Discorides praised honey as 'good for all rotten and hollow ulcers', but like many other folk remedies, honey was for a long time treated with great suspicion by the medical profession. As recently as 1976 its use was described as 'worthless but harmless' in an editorial in *Archives of Internal Medicine*. Since then, however, a good deal of individual case studies and research have supported the view that wounds treated with honey may, in many cases at least, heal more quickly than those treated with antibiotics. But it is still not clear how the honey treatment works.

Part of the effect may be due to honey being a supersaturated sugar solution. This means it contains very little water, so treating a wound with honey may help absorb moisture from the wound, denying the bacteria the water they need to survive.

Some studies, on the other hand, have shown that honey can be more effective when diluted. This seems to be because of an enzyme in honey that releases hydrogen peroxide, which is known to have an antibacterial effect. The antiseptic effect of the alkaline hydrogen peroxide occurs only when the natural acidity of honey itself is diluted and neutralized by body fluids. Yet the acidity of honey may also play a part in the fight against bacteria.

All these effects sound rather basic compared with the sophistication of antibiotics, and it seems likely that honey possesses some other healing component that has not been identified. The honey from manuka trees in New Zealand seems particularly potent as an antibacterial agent, but because we do not know what ingredient is responsible, the hypothetical molecule is simply called by the registered trade mark 'Unique Manuka Factor'.

285. Why don't the immune systems of pregnant women reject their foetuses?

Our auto-immune systems are designed to reject foreign tissues. That is the perennial problem of transplant surgery: the host body naturally rejects the transplanted organ unless drugs are given to suppress the rejection mechanism. But if the immune system reacts so violently towards foreign bodies, why do pregnant mothers not reject their foetuses? After all, half the genes of the foetus come from the father and are responsible for the creation of proteins that one would expect the mother's body to recognize as not part of her.

One theory holds that the placenta provides a barrier to keep the immune cells from getting at the foetus; another idea is that the foetus itself somehow has a way to hide the proteins it produces from the mother's immune system. Yet no one has come up with a precise idea of how either of those would work. Research with pregnant mice has supported a third idea: that the foetus itself produces an enzyme that inhibits the development of immune cells. This enzyme may create a 'no-auto-immune zone' to protect the mouse foetus from the mother's natural defence mechanism.

Whether this also happens in humans has yet to be confirmed, but it is already clear that the auto-immune system is even more sophisticated and complex than had previously been thought.

286. What molecular mechanisms are responsible for the association of obesity with certain types of cancer?

There is a great deal of evidence that obesity is linked to increased rates of certain types of cancer. One study has even claimed that 25 per cent of all cancer cases worldwide are due to excess weight and a sedentary lifestyle. But we have almost no idea why being too fat should lead to cancerous growths.

Since obesity is known to be linked to diabetes, which seems to be caused by the effect of excess fat on the body's insulin production, something similar may be contributing to the cancer link, but whether it is insulin-related, or related to some other substance, or whether fat has an effect on the mechanism that causes cancer-causing genes to come into operation, are all questions that need a good deal of further research.

287. Are mummies good for you?

From the twelfth to the eighteenth centuries, 'powdered mummy' was a common item sold by apothecaries across Europe. The Science Museum in London even has an exhibit that it describes as 'Green cylindrical wooden herb box with lid to contain mummia (powdered mummy), now empty, European, 1601–1800'.

Originally, the use of powdered mummy as a drug was inspired by the supposed curative powers of bitumen, which it was incorrectly thought was used by the Egyptians in the mummification process. Egyptian mummies were seen as a cheap source of bitumen, which was even responsible for their name: *mummiya* was a Persian term for bitumen.

At some point in the Middle Ages people got the wrong end of the mummified stick and started to believe that the supposed medicinal properties of the powder were due to its having been a living body. This belief resulted in the appearance of medicinal recipes asking for the dried body of a hanged man, or some similarly gruesome ingredient.

In the eighteenth century, physicians began to lose faith in powdered mummy and the item, once so popular, faded from apothecaries' lists. No controlled studies have ever been performed, as far as is known, to test whether powdered mummy is good for you, with or without bitumen.

See also DISEASE 119–31

MEMORY

288. How and where are memories stored in the brain?

In the 1920s the American psychologist Karl Lashley conducted a famous experiment to determine where memories are stored in a rat's brain. After teaching the rat to run through a maze, he cut out a bit of its brain to see if he could excise the memory. But as long as the rat was still functioning, it always seemed to remember what it had learnt, whichever part of the brain he cut out. His conclusion was that memories are not stored in any particular place, but are spread around all areas of the brain.

According to current theory, memories are created in the hippocampus and later transferred to the frontal lobes, but we still need the frontal cortex and the hippocampus to retrieve them. The whole business of laying down a memory, putting it away somewhere safe and retrieving it at a later date is clearly extraordinarily complex, and made all the more so by the realization that every time you recall a memory, you create a new memory of having recalled it. So whether a memory exists in a particular place in the brain seems almost as doubtful as ashley thought, though one cannot help feeling that it must be somewhere.

289. What form do memories take in the brain?

In the 1960s, when the recent discoveries of DNA and RNA were seen as the potential answers to everything we needed to know about life, there was a theory that each new memory corresponded to the creation of a new RNA molecule. Among the experiments designed to test the theory were various attempts to extract memories from the brain of a rat or goldfish and transfer them physically to another individual of the same species. There was even one famous experiment in which a flatworm was taught something, then minced and fed to another worm to see if knowledge was edible. Even though the results of some of these experiments seemed to confirm the RNA theory of memory, the whole idea was soon quietly abandoned, though recent research suggests it may not have been totally misguided. Memories are now seen as having something to do with the creation of new synapses joining certain neurons (nerve cells) in the brain, or the strengthening of existing synapses. Such synaptic alterations require the creation of new RNA and proteins in the hippocampus, and experiments with mice have shown that when RNA synthesis is stopped, it has an adverse effect on the subject's ability to consolidate memories.

But we still don't know what memories look like.

290. What is the difference between the way the brain stores long-term and short-term memories?

The division of memory into short-term and long-term has been with us since the nineteenth century, and offers a very plausible model for the way we experience memory. The idea that thoughts are held for a brief time in short-term memory while we decide whether to retain them in long-term memory or let them fade away is enticing. Yet opinions are divided on whether there really are two different types of memory – and if there are, what the difference is. Attempts to identify a boundary between the two types of memory have been inconclusive.

What you don't know would make a great book.

Rev. Sydney Smith (1771–1845)

291. By what mechanism do we recall past memories?

In the 1960s researchers performed an intriguing experiment on goldfish. The fish was placed in a T-shaped tank and taught to swim up to the crossbar, then turn in a particular direction. This is simply accomplished by placing food at one or other end of the crossbar. Before the experiment begins, however, some alcohol is added to the water.

When the goldfish, by now a bit tipsy, has learnt which way to turn, it is sobered up by being placed in clean water. Further experiments show that it is now liable to have forgotten what it learnt. But put it back into alcoholic water and it will remember it again. Humans have shown a similar tendency: learn something when you are drunk and you may forget it when you sober up. Next time you get drunk, however, it may well all come back to you.

These are examples of memory being context-dependent, which supports a theory of memory recall known as 'encoding specificity'. A memory, under this theory, is not stored in isolation but together with the situation in which it was formed.

The other main theory is that recalling a memory is a two-stage process, first involving search and retrieval, then recognition. That at least would explain why recognizing something you have seen before is a much easier process than recalling what it looked like: recognition only involves one of the two stages of recall.

In either case, the mechanism by which it all happens has yet to be discovered.

292. Is consciousness a high-level consequence of memory in all its forms?

Consciousness involves being aware that I exist and knowing who I am, or at least having some sense of identity. You could say that our sense of identity is nothing more than the total of all our memories, and the knowledge that we exist is provided by the constant flow of experiences and sensations into our short-term memory. So you could say that consciousness is a direct consequence of all our memories.

On the other hand, you could say that computers have memories, but don't have consciousness. Perhaps consciousness is what gives human memory its unique quality.

See also CONSCIOUSNESS 101–5

293. Is there any hope of discovering how memories work with our current understanding of physics, or does an entire new level of physics need to be discovered?

As was mentioned in the introduction to the section on the brain, there has been a tendency throughout the history of science to explain the brain's functioning by analogy with the latest technology, whether that is plumbing or electronic computers. In his 1989 book *The Emperor's New Mind*, Sir Roger Penrose suggests that certain aspects of our mental functioning – notably the problem of consciousness – may not be explicable by our current understanding of physics. When we cannot say how memories are stored, where they are stored, how they are retrieved or what form they take, it strongly suggests that the real problem is a lack of understanding of the physical processes involved in the brain. Penrose proposes a hypothesis in which quantum mechanics somehow plays a role. However, some philosophers argue that memory and consciousness are by their very nature subjective, and thus not entirely amenable to mechanistic explanations.

294. Why can we not remember anything of our earliest years, when it is clear that babies have good working memories?

What is your earliest memory? For most people, the first thing they can remember dates from the age of about three, but babies can remember faces and have an amazing capacity to learn things, which involves having a memory. So what happens to all the memories of our infant years?

Could it be that the acquisition of language alters the way we remember things, almost as though our brains have been reprogrammed with a new operating system that can no longer consciously access the older memories of our infant years? And if we do remember things from infancy, how can we tell whether we are remembering them or remembering what we were told about them by our mothers, or remembering a misremembered version of it from our slightly later years?

295. Is there truly any such thing as a 'memory', or is it just a convenient human construct, a catch-all term for a wide variety of mental activities that we do not understand?

I can remember that the Battle of Hastings was in 1066; I can remember what I had for breakfast this morning; I can remember seeing the Northern Lights in Iceland thirty-five years ago. Yet these memories are of very different types: the first is a firm memory of a fact, and the other two are personal experiences, one much more recent than the other. My memory of breakfast is probably fairly accurate, as it was not very long ago and has not had the opportunity to fade or change, but my memory of the Northern Lights has been accessed many times and probably changed a little every time. It is not so much a single 'memory' as the amalgamation of all the memories I have had of that event over several decades all merged and blurred into a single entity.

Studies of people's recollections of their youth have shown that such 'memories' can be wildly at variance with the historical facts. You think you are remembering something, but what you recall is a reconstruction of an embellishment of a version of something you thought you remembered. And that, more often than we would like to admit, is what a 'memory' really is.

THE MIDDLE AGES

296. Why did the Danes turn into Viking pirates?

From the late eighth century, the previously stay-at-home people of Scandinavia embarked on a period of dramatic expansion – raiding, looting, colonizing and trading from Russia and Constantinople in the east, to Spain and North Africa in the south and Iceland and Greenland in the west. They even reached as far as Newfoundland and Labrador in North America, centuries before Cabot and Columbus rediscovered the New World.

What brought about this huge change in culture? There are three main theories:

(i) It was all prompted by the desire of Charlemagne, king of the Franks from 768 and emperor of the Romans from 800, to impose Christianity by force. The battle between the forces of Christendom and Norse paganism had already been raging for almost a century, so the Viking expansion may have been motivated by a desire to protect the pagan Norse culture and to wreak revenge on Charlemagne.

(ii) It was all caused by a population boom that left the Norse people with insufficient agricultural land to feed their growing numbers. With their nautical skills already highly developed, taking to the high seas was an easier option than clearing the forests to the east.

(iii) There was a long-term European economic slump caused by the fall of the Roman Empire and the rise of Islam, both of which events had led to a decline in international trade. The Viking expansion was driven by a need to open up new markets.

None of these sets of circumstances completely accounts for the vast extent of the Viking expansion, but any of them might have planted the seeds for an era of successful belligerence.

297. Who invented firearms?

Gunpowder is thought to have been invented by the Chinese in the ninth century, possibly by Daoist monks looking for an elixir of eternal life. Its use in weaponry, particularly firebombs, began shortly afterwards, but the earliest account of firearms – guns – dates from the twelfth century. A sculpture from that period found in a cave in Sichuan shows a man carrying a type of primitive cannon known as a bombard, with flames and a cannonball emerging from its mouth. The oldest surviving cannon of this type has been dated to 1288. The unanswered question is who first had the idea of using the explosive power of gunpowder to propel metallic projectiles at an enemy. All we know is that he was probably Chinese, lived around the twelfth century, and invented something that led to the arms industry as we know it today.

298. How many Normans invaded England in 1066?

For one of the most significant events in British history, we have remarkably few precise details about the Battle of Hastings. Perhaps that is not surprising when our main source is a lengthy embroidered cartoon (the Bayeux Tapestry), which was probably not completed until eleven years later. Even the story about King Harold being killed by an arrow in his eye is very much disputed.

As to the size of the armies, the English are thought to have numbered 7,000–8,000, but there are no reliable sources for the size of the army of William the Conqueror. Most writers believe it was about the same size as the English army, but attempts to give a more precise figure have varied between 4,000 and 20,000.

299. Did the so-called Malfosse Incident really happen during the Norman Conquest, and if so, where?

At least five chroniclers of the eleventh and twelfth centuries mention an incident that supposedly took place at a site known as Malfosse ('evil ditch'). Their accounts differ in various ways, but tell a similar tale of the Norman pursuit of desperate English troops in the closing stages of the Battle of Hastings.

The English have decided to make a last stand at Malfosse where a ditch, or some other natural formation, offers them an advantage. The Normans are led by Eustace of Boulogne and William himself, and the two are conferring. An English soldier, either wounded or playing dead on the battlefield, sees the two French knights deep in discussion. He gets to his feet and hits one of the knights with a stone between the shoulder-blades with such force that blood pours out of his mouth and nose and he is carried away badly wounded.

French accounts of the battle omit any reference to the incident, which if it happened would have been a narrow escape for William. At least five possible sites have been suggested for Malfosse, but nothing has been found to confirm the incident.

300. Where was the Bayeux Tapestry made?

The Bayeux Tapestry, which is not a tapestry at all but a piece of embroidery, is thought to have been completed in 1077, and is our main source of information about the Battle of Hastings eleven years earlier. But who commissioned the Bayeux Tapestry, who did the embroidery and where it was made have long been disputed.

French tradition holds that it was commissioned by Queen Matilda, wife of William the Conqueror, and made by herself and her ladies in waiting. Others say that it was commissioned by Bishop Odo, William's half-brother, which might help explain why it was found in Bayeux Cathedral in Normandy, which was built for Odo.

Experts in historical embroidery, however, have suggested that the style and expertise of the Tapestry suggest English not French origins, and the style of lettering in the Latin words it displays again point to England as the source. In particular, the letter U always appears as V, which was much more the English style.

Since the earliest known reference to the Tapestry is in an inventory of Bayeux Cathedral from 1476, the first four hundred years of its history are lost. Until that gap is filled we will have to continue to guess where the elaborate stitching began.

301. Where is Genghis Khan's grave?

When the great Mongol emperor Genghis Khan died in 1227 at the age of about sixty-five, he was buried, according to tradition and his expressed wishes, in an unmarked grave in a secret location. So secret, in fact, that according to legend his funeral escort killed anyone they encountered, including the slaves who built the tomb, and on their return were themselves killed. Some say that a river was diverted over the site, while another story is that horses trampled the land and trees were planted to hide the grave. Whatever the secrecy measures were, they certainly seem to have worked, for the site of Genghis Khan's grave is still unknown.

While we are on the subject, the manner of his death is also the subject of various different stories. One such tale concerns a princess captured as war booty from battles with the Tangut in China. The princess is said to have concealed a pair of pliers in her vagina, and these damaged Genghis Khan so badly that he died of the injuries. That might at least explain why he wanted his grave kept secret, but the tale is generally held to have been invented by his Mongolian rivals.

302. Did Marco Polo reach China, or was his account of that land all made up?

The Venetian merchant Marco Polo (1254–1324) was undoubtedly one of the great travellers of the Middle Ages. His book *Il Milione* (known in English as *The Travels of Marco Polo*) was a medieval best-seller, translated into many languages, yet there are some aspects of his account that suggest that he may not have travelled as far as he claimed. In particular, his writings about China have been cited as evidence that he was merely recounting tales told to him by others, rather than narrating his personal experiences.

For example, his account, while detailed in some respects, fails to mention Chinese calligraphy, or chopsticks, or tea, or foot binding, or even the Great Wall. What is more, no mention of Marco Polo has been found in Chinese records of the time, despite his claim to have served as a special emissary for Kublai Khan, the Chinese emperor. All the same, Marco Polo was the first European to mention tigers, the imperial Chinese postal system, Japan, and the Grand Canal linking Beijing to Hangzhou.

On his deathbed, Marco Polo is said to have been urged by a priest to confess that he had made up much of his book. In reply, he is reputed to have said, 'I have not told half of what I have seen.'

303. What was the cause of the Black Death?

In the middle of the fourteenth century, the Black Death is thought to have killed around half of Europe's population and reduced the world population by about 20 per cent. Yet we cannot be sure what caused it, or even what it was. Until recently it was generally assumed that the Black Death was a form of bubonic plague, similar to that which caused epidemics in European cities in the seventeenth century. But since the beginning of the twenty-first century, following studies of detailed contemporary accounts of the Black Death, doubts have been expressed as to whether it really was bubonic plague. While the symptoms were in many ways similar, the Black Death struck at different times of the year, and had a different fatality rate and differing levels of recurrence and periods between outbreaks, from those experienced during the later epidemics of bubonic plague. Its contrasting behaviour in cities and rural communities also seemed to differ from the behaviour of bubonic plague. This led several researchers to suggest alternatives, such as a form of haemorrhagic fever (similar to ebola) or even anthrax, as the main cause of the Black Death.

In 2010, however, a detailed analysis of the DNA found in the teeth of bodies found in medieval plague pits found evidence of a bacterium associated with bubonic plague. If that is true, there is still the question of how the disease was transmitted to humans. We know that bubonic plague can be carried by the fleas that live on rats. Outbreaks of the Black Death, however, are known to have occurred in areas too hot for fleas to survive – but that is also true of nineteenth-century epidemics of bubonic plague in China, India and the Middle East. Yet the pustules or buboes of nineteenth-century victims of bubonic plague appeared mainly in the groin, where fleas are known to bite. In the case of the Black Death, the buboes frequently appeared also on the armpits and neck, which are distinctly less popular feeding grounds for fleas. Finally, it had been suggested that outbreaks of bubonic plague were linked to a decrease in the rat population, which forces the fleas to jump to humans to find other feeding grounds. No such changes in the rat population have been identified in connection with the Black Death.

304. Who was Perkin Warbeck?

Perkin Warbeck (1474–99), pretender to the throne of England, is one of the most mysterious characters in English history. Claiming to be Richard of Shrewsbury, Duke of York – the younger son of Edward IV, who was believed to have died in the Tower of London

(→ MURDER 321) – he made several attempts to seize the throne from Henry VII. He was officially recognized as Richard of Shrewsbury by Margaret of York, sister of Edward IV, though it is unclear whether this was because she believed him or because she saw him as a useful political ally.

When he was captured by the king's forces, he made a confession in which he gave details of his Flemish origins, and claimed to have learnt English in Ireland. As this confession was given under duress, it is considered nothing more than a fabrication concocted in an attempt to avoid the death penalty.

In appearance, Perkin Warbeck was said to have resembled Edward IV, leading some to speculate that he may have been the king's illegitimate son, or even the prince he had claimed to be. Perkin Warbeck was hanged in 1499 and his body sent for dissection. As nothing of him remains, even DNA testing cannot solve this mystery.

See also ENGLISH HISTORY 169–172, MURDER 317–19, 321, POPES 387–8

MODERN HISTORY

305. Who is the soldier buried in the Tomb of the Unknown Warrior in Westminster Abbey?

There is only one tombstone in London's Westminster Abbey on which it is forbidden to walk. That stone bears a plaque with a long inscription beginning with the words:

BENEATH THIS STONE RESTS THE BODY OF A BRITISH
WARRIOR UNKNOWN BY NAME OR RANK BROUGHT
FROM FRANCE TO LIE AMONG THE MOST ILLUSTRIOUS
OF THE LAND.

This is the tomb of the Unknown Warrior.

The idea for such a memorial began in 1916, when an army
chaplain, the Reverend David Railton, saw a grave in France marked
with a wooden cross and the words 'An Unknown British Soldier'.
Railton wrote to the Dean of Westminster in 1920 proposing that
an unidentified British soldier, killed in the war that had just
finished, should be brought from France and given a state funeral
in Westminster Abbey.

The idea was approved and a selection of unidentified bodies
were exhumed from various battlefields and brought to a chapel in
France. Each was on a stretcher and was covered by a British flag. A
British brigadier general then chose one of the bodies, not knowing
which battlefield any of them had come from, and that body was
brought to London to be buried as the Unknown Warrior. The other
bodies were taken away for reburial.

306. How did Amelia Earhart die?

On 2 July 1937 the pioneering American pilot Amelia Earhart went
missing on a flight over the Pacific. Neither she nor her navigator
Fred Noonan was ever seen again, and their plane has never been
found. She was declared legally dead on 5 January 1939. Becoming
the first woman to fly solo across the Atlantic in 1932 had earned
Earhart unique celebrity status around the world, and news of her
disappearance was met with grief and a determination to find out
what had happened. All that we are left with, however, are a few pieces
of disputed evidence and a large number of speculative theories.

The most likely explanation is that her plane, a Lockheed Electra,
ran out of fuel and her attempt to land it on the small airstrip on

tiny, remote Howland Island, near where her last message came from, failed owing to faulty navigational equipment. The plane crashed into the sea and Earhart and Noonan drowned.

Another popular theory is that they crashed not in the sea off Howland Island but about 350 miles away on Gardner Island (now Nikumaroro in Kiribati). In 1940 a British colonial officer found a skeleton on Gardner Island which he believed to be that of a woman, together with an item of aircraft equipment. The bones were sent to Fiji where, after an examination, they were declared to be those of a short male. A more recent examination of the data, however, concluded that they were from a tall white female of Northern European ancestry. Unfortunately this was based only on bone lengths, as the bones themselves had long before gone missing in Fiji.

In 2010 another group of researchers found what they thought were bones from a human finger on Nikumaroro. DNA analysis was inconclusive, but suggested that they might be from a human or a sea turtle. Other artefacts have been found on Nikumaroro, which may have come from a plane similar to Earhart's Electra.

A rather more outlandish theory is that Earhart and Noonan crashed on an island occupied by the Japanese and were executed as spies. Some take this theory seriously, despite the fact that it is similar to the plot of a 1943 film called *Flight for Freedom*, in which a character reminiscent of Earhart undertakes a spying mission against the Japanese.

307. Did Hitler ever issue written orders for the 'Final Solution' – the extermination of the Jews of Europe?

The 'Final Solution' seems to have been such a clear and universally accepted plank of Hitler's policy that it seems unthinkable that no such order was given, yet no documentary proof has ever been unearthed. The Nazi propaganda minister Joseph Goebbels alludes in his diary to the Führer making his objectives clear on the matter, but despite a great deal of research, no written orders have been found. Perhaps none were given; perhaps all traces of them were destroyed; perhaps they will turn up tomorrow.

308. What happened to the Amber Room?

In 1716 King Wilhelm I of Prussia gave a fabulous gift to his ally, Tsar Peter the Great of Russia. Built between 1701 and 1709, it was a complete room decorated with amber panels backed with mirrors and gold leaf. Known as the Amber Room, it was eventually set up in the Catherine Palace near St Petersburg, and covered an area of some 55 square metres (600 sq ft).

The Amber Room remained much admired and generally undisturbed until the German invasion of the Soviet Union in 1941. The Russians tried to disassemble it but the material proved too fragile. The Germans, however, managed to do so safely, and in October 1941 the Amber Room was brought to Königsberg Castle in Prussia in twenty-seven crates.

Early in 1945, crates were again seen leaving Königsberg Castle, but no information exists on what was in them. Later the city of Königsberg (now Kaliningrad in the Russian Federation) was heavily bombed by the British and the castle was destroyed by the Russians. The Amber Room has never been seen again, though rumours of the discovery of hiding places or signs of gold and/or amber continue to surface every few years.

309. Who was the man whose body was washed up on Somerton Beach in Australia in 1948 in the so-called Tamam Shud case?

At 6.30 a.m. on 1 December 1948, a man was found dead at Somerton Beach in Adelaide, Australia. Apart from being dead, he appeared to be in fine condition and aged about forty-five. Close examination revealed that all identifying labels had been removed from his clothes. Neither his fingerprints nor dental records matched anything in Australia, and when the search was extended worldwide, it didn't help. An autopsy revealed internal abnormalities, including a grossly extended spleen, which were thought to be consistent with poisoning, but no traces of any known poison were found.

An appeal for possible sightings of the man in Adelaide before his death led to the discovery of a brown suitcase that contained more clothes with labels removed and a jacket that had been made in America. A hidden pocket was found stitched into his trousers which contained a piece of paper apparently torn from a book containing the words 'tamam shud'.

Thanks to good literary detective work, the words were recognized as a Persian phrase that appears at the end of Edward Fitzgerald's translation of the *Rubáiyát* of Omar Khayyám. The phrase means 'ended'. The very book the words had been torn from was later found, thrown on the back seat of an unlocked car shortly before the man's death. The book was examined and found to be a very rare edition of the work. It also contained what looked like a coded message, which nobody has yet been able to crack. All attempts to identify the man have also come to nothing.

310. What proportion of all the people who have ever lived are alive today?

It is frequently claimed by lovers of useless information that half the people who have ever lived are alive today. That is complete nonsense, but even estimating the correct figure is not easy. Before the middle of the nineteenth century, accurate population figures were not known for large areas of the world. We know that the population is now approaching 7 billion; it reached 6 billion in 1999, 5 billion in 1987, and is estimated to have first passed the billion mark in 1804.

Around the birth of Christ, the world population is thought to have been about 200 million, but before that everything is guesswork; and even if we only go back as far as the evolution of modern man, *Homo sapiens sapiens*, we have at least 200,000 years of humans to count.

Making certain assumptions about population growth, and assuming there was no great population explosion or disaster we do not know about, then it has been estimated that the total number of humans who have ever lived is about 110 billion, of whom just over 6 per cent are alive today. But we will probably never know the true figure.

THE MOON

311. Did the Moon split from Earth after a massive collision?

We know that the Moon has remained relatively unchanged over the past 2.5 billion years, but how it came into being is still not known for sure. There are four theories concerning its origin.

(i) It may have formed from coalescing gas and dust at the same time as the Earth.

(ii) It may have been an asteroid that came close to Earth and was captured by the Earth's gravitational pull.

(iii) It may have spun off from Earth during the early period of our planet's formation.

(iv) It may have been formed from rock and debris from Earth that was knocked off in a collision with another planet about the size of Mars.

Generally, the last of these theories is thought to be the most probable, but we cannot be certain. It would at least account for the fact that the Moon always has the same side facing Earth: its pattern of rotation would have been inherited from that of its mother planet.

312. What, if anything, caused the Moon's Linné Crater to change its appearance in the mid-nineteenth century?

Among the features on the surface of the Moon is a beautiful bowl-shaped crater that is named after the great Swedish zoologist and botanist Carl Linnaeus. The Linné Crater, which is only about 10 million years old (very young for such a crater), began to attract considerable attention in 1866 when the German astronomer Johann Schmidt claimed to have detected a change in its appearance.

Schmidt, the director of the Athens Observatory, was a highly respected scientist who had a particular interest in and expertise regarding the Moon, of which he was preparing a map that was better than anything previously seen. For several decades, great interest centred on the shape-changing crater, though the general conclusion was that Schmidt's observation was caused by his optical instruments rather than the crater itself.

More recently, further changes have been observed at Linné, and these have been put down to 'residual outgassing' (i.e. the emission of a gas), which could distort the appearance of the crater. Schmidt's 1866 observations could have been caused by an unusual increase in the amount of gas produced. So something may be going on at Linné that needs explaining after all.

See also WATER 471

'Beauty is truth, truth beauty,' – that is all
Ye know on earth, and all ye need to know.

John Keats (1795–1821)

MOZART

313. What was the cause of Mozart's death?

Wolfgang Amadeus Mozart suffered from so many illnesses during his lifetime (1756–91) that it is difficult to tell what killed him at the tender age of thirty-five. According to his death certificate, he died from 'severe miliary fever', but 'miliary' just means small bumps on the skin, so is more a description of the symptoms than a diagnosis of the disease.

Mozart's wife Constanza is said to have spoken of his suspicions that he had been poisoned, which contributed to the persistent suggestion that the Italian composer Antonio Salieri was responsible for his rival's death (a rumour that inspired a verse drama by Alexander Pushkin, an opera by Nikolai Rimsky-Korsakov and Peter Shaffer's play and film *Amadeus*). But according to the medical evidence, poisoning is unlikely. If Mozart was poisoned, the most likely suspect is not Salieri but the proprietary medicines containing antimony that Mozart took for other real and supposed illnesses. He was quite a hypochondriac.

One recent study concluded that the most likely diagnosis was rheumatic fever. Another declared that it was probably a streptococcal infection known locally as *Wassersucht* leading to an acute nephritic syndrome caused by post-streptococcal glomerulonephritis. Or kidney failure, as you and I would call it.

314. How much of Mozart's *Requiem* did Mozart actually write?

The anonymous stranger who commissioned Mozart's *Requiem* towards the end of the composer's life is now known to have been Count Franz von Walsegg, who is thought to have adopted the subterfuge in order to pass off the work as his own. When Mozart died with the *Requiem* unfinished, it was completed by his friend and colleague Franz Xaver Süssmayr and delivered to Walsegg, but how much of the final score was genuine Mozart and how much is attributable to Süssmayr has never been satisfactorily sorted out.

The original autograph manuscript clearly shows the Introit in Mozart's hand, together with much of the Kyrie and Dies Irae, and the vocal parts of the offertory. Süssmayr spoke of 'scraps of paper', now lost, for much of the remainder of the composition, though he later claimed that the Sanctus and Agnus Dei were all his own work.

Ten years younger than Mozart, Süssmayr may have been his student and certainly worked as his copyist. He went on to a reasonably successful career as a composer himself, particularly of church music and opera, but never produced anything remotely as impressive as Mozart's *Requiem*.

315. Is the skull displayed at the International Mozarteum Foundation in Salzburg really that of Mozart?

In 1902 the International Mozarteum Foundation in Salzburg was presented with an intriguing if somewhat macabre gift: Mozart's skull. Or that at least was what it purported to be, and the story it came with provided some corroboration.

The usual romanticized account of Mozart's body being tossed into a paupers' mass grave on a stormy night has not stood up to examination, and the accepted view is that he was buried in a wooden coffin in a grave with four or five other people, which was the normal middle-class custom in Vienna at the time. A few years later, the bones were dug up to enable the plot to be used by others, which was also normal practice. The story behind the Mozarteum skull is based on an account of Joseph Rothmayer, one of the original gravediggers, who said that he attached a wire round the neck of Mozart's body so that he would be able to identify it. When the bodies were dug up in 1801, Rothmayer said he took the skull away with him, and that was the skull presented a century later to the Mozarteum.

In 2006, to celebrate the 250th anniversary of Mozart's birth, DNA studies were commissioned to compare the bones in the skull with samples taken from thigh bones of two skeletons in the Mozart family grave. Not only did these analyses not solve the problem, they even introduced another one. For not only was Mozart's skull found to show no relationship to the others, but the others showed no relationship to each other. At least one of them must therefore have not been a Mozart, raising the question of the identity of the uninvited corpse in the family vault. The skull, in fact, may be the only true Mozart in the trio. The DNA test was therefore ruled inconclusive.

MURDER

316. Was Pharaoh Ramses III assassinated?

Ramses III ruled Egypt for thirty-one years around 1170 BC and has been described as the last of the great pharaohs. But he ruled at a time of great turmoil, beset by wars and economic difficulties –

indeed, his inability to pay his workmen led to the earliest general strike in known history. Yet that was not the reason for the plot to assassinate him.

That such a plot existed is made clear by a set of trial documents that led to thirty-eight people being sentenced to death. It all began with a squabble between two of his wives over which of their sons would succeed Ramses, but it developed into a major conspiracy involving state officials, army personnel, scribes and even members of the royal harem.

Ramses died before the sentences were carried out, but whether the assassination plot had succeeded is unclear. His mummy shows no obvious wounds, and no traces of poison or snakebite were found, both of which have been suggested as causes of his death. However, his mummy was equipped with an amulet to protect him from snakes in the afterlife.

317. Who killed Edward the Martyr in 978?

Edward the Martyr had a short but troubled life. The eldest son of King Edgar, he came to the throne of England in 975 when he was only about thirteen years old. His reign began with the inauspicious sighting of a comet, continued with a famine, and ended three years after it had begun when he was murdered at Corfe Castle in Dorset. According to the *Anglo-Saxon Chronicle*, he was buried without honours or ceremony.

Some say his murder was organized by his step-mother Ælfthryth, whom he was visiting at Corfe; some say the man behind it – who was also at Corfe – was Edward's younger half-brother Ethelred (the Unready – but ready enough to take the throne when Edward was dead); some say that Edward's uncontrolled rages had upset so many people that anyone could have killed him. The version given by William of Malmesbury, writing two centuries later, asserts that Ælfthryth.

allured him to her with female blandishment and made him
lean forward, and after saluting him while he was eagerly
drinking from the cup which had been presented, the dagger of an
attendant pierced him through.

William's contemporary, Henry of Huntingdon, says that it was Ælfthryth who wielded the knife. But William and Henry were writing at such a remove from the events that they describe that they cannot be judged reliable.

318. Was Godwin, father of King Harold II, poisoned?

Thanks to his political skills, Godwin, Earl of Wessex, was probably the most powerful man in England when Edward the Confessor came to the throne in 1042. Even though Godwin was widely held responsible for the murder of Edward's brother, Alfred the Atheling, Edward was content to marry Godwin's daughter to cement his ties with the powerful earl. But since Edward had taken a vow of celibacy, nothing much came of the union. No children, anyway.

As usual for any medieval power behind the throne, Godwin accumulated enemies, and his situation was not helped when his son Swegen was outlawed in 1046 for seducing the Abbess of Leominister. Relations with Edward deteriorated, Godwin even at one stage raising an army against the king. That particular dispute was settled without bloodshed, but when Godwin died at a banquet at Winchester in 1053, there were two very different accounts of what happened.

The *Anglo-Saxon Chronicle* relates how Godwin suddenly sank against his footstool, speechless and helpless. He was carried into the king's chamber, where he lingered on for several days, still immobile and unable to speak, until he finally died.

The Norman version, however, written long after the event, tells of Godwin holding up a piece of bread and denying to the king that he had had a hand in his brother's death: 'God forbid that I should swallow this morsel, if I have done anything which might tend either to his danger or your disadvantage.' He then popped the bread – which in some versions of the story is a communion wafer – into his mouth and choked to death.

Suspicious circumstances, one might say, in either case.

319. Was William Rufus murdered?

On 2 August 1100 King William II of England – known as William Rufus, probably because of his ruddy complexion – went out hunting in the New Forest and was shot through a lung with an arrow. His companions said it was just a hunting accident, made worse when the king fell onto the arrow after the shot, driving it further into his body. He died shortly afterwards, and his body was left on the spot as those who had been with him fled to make preparations for the succession.

The main suspect – if it was not an accident – has always been Walter Tirel, possibly acting for William's brother Henry, who succeeded him as Henry I. But no charges were ever laid, let alone any conviction achieved. Yet even the accounts that say it was an accident (for example, William of Malmesbury writing c.1125 and Orderic Vitalis c.1135) say that it was Tirel's arrow that killed the king. William Rufus did, however, have many enemies, notably in the Church, especially after Anselm, Archbishop of Canterbury, had been forced into exile.

320. Was Agnès Sorel, mistress of Charles VII of France, murdered?

Agnès Sorel (1421–50) was Charles VII's favourite mistress. She bore him three daughters and was pregnant with the fourth when she died suddenly at the age of twenty-eight. The immediate diagnosis was dysentery, but a forensic examination of her disinterred remains in 2005 revealed that the cause of death was mercury poisoning.

In the fifteenth century, mercury was commonly used in cosmetic preparations or to treat certain diseases, so an innocent explanation is not ruled out. Both Charles's son, the future Louis XI, and one of his ministers, Jacques Coeur, might have stood to gain from Agnès Sorel's removal, but there is no direct evidence against either.

Charles VII himself was evidently not too inconvenienced by her death, as he swiftly recruited her cousin, Antoinette de Maignelais, to take over her role in his bed.

321. Were the Princes in the Tower murdered, and if so, by whom?

When Edward IV of England died in April 1483, his brother Richard, Duke of Gloucester, was named 'Lord Protector of the Realm' for the late king's son and obvious successor, the twelve-year-old Edward V. Eager to assume power, Richard had both the royal princes – Edward and his younger brother Richard of Shrewsbury (→ THE MIDDLE AGES 304) – declared illegitimate by Parliament on the grounds of doubts about the legality of the marriage between Edward IV and their mother, Elizabeth Woodville.

In the run-up to Edward's coronation, the young princes were staying at the Tower of London, then simply a royal residence. They were seen playing in the grounds in the summer of 1483, but were never seen again. In 1674, during renovations at the Tower, the skeletons of two children were found under a staircase, but no identification was possible and they were reburied. In 1933 the grave was again opened, but once more, no identification was possible.

Next time, perhaps they will use DNA testing – but there are no plans for another exhumation.

322. Who murdered Jean-Marie Leclair?

Jean-Marie Leclair the Elder (1694–1764) was a renowned violin virtuoso and one of the leading French composers of his time, specializing in works for the violin, though he also wrote one opera. He was also employed as premier dancer and ballet master in Turin and, moving between that city and Paris, he was celebrated as the most travelled French musician of his age.

Shortly after the break-up of his second marriage, Leclair bought a small house in Paris, and it was here that he was found dead in October 1764. He had been stabbed in the back. Suspicion fell on his ex-wife and his nephew, but no evidence was ever found to implicate them, or a rival musician, which seemed the only other likely possibility. The murderer or murderess was never discovered.

323. Who killed Mary Rogers, whose death inspired Edgar Allan Poe to write his second detective story?

In 1841 Edgar Allen Poe (→ WRITERS 486–7) wrote 'The Murders in the Rue Morgue', which is often acknowledged as the first detective story. The following year he wrote his second detective story, 'The Mystery of Marie Rogêt', but this time there was a big difference: it was very closely based on a real event.

Mary Rogers's body had been found floating in the Hudson River on 28 July 1841, bearing signs of having been severely battered. Finger-sized marks on her neck were taken as an indication that she had been strangled. She was about twenty-one years old.

Before her death she had been employed in a tobacconist's shop in New York – largely because her great beauty was such a draw for customers. Among those who purchased their smoking requisites from the 'Beautiful Cigar Girl' (as she became known) were the writers James Fenimore Cooper, Washington Irving and, according to some accounts, Edgar Allan Poe himself.

Mary's death was seized upon by the newspapers as a sensational crime mystery. Over the following months several arrests were made, but all suspects were soon released. Increasingly bizarre stories began to circulate: that she had been killed by her fiancé; that she had died as the result of an unsuccessful abortion; that the body was not that of the Beautiful Cigar Girl at all. One later writer even named Edgar Allan Poe as a possible suspect. The case has never been solved.

324. Who was Jack the Ripper?

Between August and November 1888 five women were violently killed in London's East End. The bodies were slashed and mutilated. A further six murders, both before and after this period, were linked to those five, though differences in method raised doubts that they were by the same killer. Nearly all the victims were prostitutes. A letter to the police, purportedly from the murderer but probably a fake, was signed 'Jack the Ripper'. The name stuck.

The case has fascinated both professional and amateur sleuths ever since, and over a hundred potential suspects have been named, ranging from the writer Lewis Carroll and the painter Walter Sickert to Prince Albert Victor, Duke of Clarence and son of the future King Edward VII. Yet all we know is that the killer may have possessed medical knowledge and may have been left-handed. Even those facts, however, are disputed.

325. Who killed Lizzie Borden's parents?

An otherwise unremarkable New England spinster has been immortalized in a rhyme chanted by generations of American girls as they spin a skipping rope for their playmates to jump over:

> *Lizzie Borden took an axe,*
> *Gave her mother forty whacks.*
> *When she saw what she had done*
> *She gave her father forty-one.*

In fact Lizzie Borden was acquitted of the hatchet murders of her father and stepmother on 4 August 1892 in Fall River, Massachusetts.

The jury reached their verdict after only an hour and a half's deliberation. But if Lizzie Borden did not deliver the fatal whacks, who did? Later writers have speculated that the maid or Lizzie's half-brother may have had motives to kill the Bordens, but neither was ever charged or apparently even suspected at the time. Lizzie died in 1927, aged sixty-six and worth around a million dollars. She left $500 in perpetual trust for the care of her father's grave.

326. Who killed Edwin Drood?

Edwin Drood is engaged to Rosa Bud, but Rosa is also loved by her choirmaster, John Jasper, who is Drood's uncle. Then Neville Landless arrives from Ceylon and also falls in love with Rosa. He takes an intense dislike to Edwin, who disappears under mysterious circumstances.

So, in the most cramped of nutshells, runs the plot of Charles Dickens's final novel, *The Mystery of Edwin Drood*, left incomplete at his death in 1870. The trouble is that Dickens never got round to revealing who killed Edwin Drood, or what role was played in his disappearance by the mysterious stranger, Dick Datchery.

Various attempts have been made to finish the story, including one by an American spiritualist who claimed to be in touch with the spirit of Dickens himself. In 1914 the Dickens Fellowship in London staged a trial of John Jasper, who had always been considered the most likely murderer, with G.K. Chesterton as the judge. When the jury returned a verdict of manslaughter, Chesterton ruled that the mystery of Edwin Drood was insoluble and fined everyone, except himself, for contempt of court.

> When ignorance gets started it knows no bounds.
>
> Will Rogers (1879–1935)

MUSIC

327. Who composed the song 'Greensleeves'?

There has long been a rumour that Henry VIII composed 'Greensleeves', but where this idea came from is anyone's guess. The style of the music was not known in England at the time, and although the king was an accomplished musician and is known to have written many songs, none of his known work is similar to 'Greensleeves'.

A ballad named 'Green Sleeves' was registered at the London Stationers' Company in 1580, and the song was included in a book dated 1584, in which it is described as a 'new tune'. Since Henry VIII died in 1547, his involvement seems unlikely.

While we are on the subject of anachronisms, it is perhaps worth mentioning that Shakespeare, in *The Merry Wives of Windsor*, has Falstaff exclaim: 'Let the sky rain potatoes! Let it thunder to the tune of "Greensleeves"!' The play is set around 1400, long before potatoes were brought from the New World to England.

328. Who wrote the music and who wrote the words for the British national anthem?

'God Save the Queen' (or 'King') has never formally been adopted as the British national anthem, but has filled that role in the life of the nation since 1745, when it was first performed in public at the Theatre Royal in Drury Lane. Both the tune and the words, however, are much older than that. The music has been attributed to England's greatest composer, Henry Purcell (1659–95), and the French composer Jean-Baptiste Lully (1632–87), among others. Another strong candidate is an earlier English composer, John Bull, who wrote a keyboard piece strikingly similar to the anthem in 1619.

The words are even older, but may have come together from various traditional sources. The phrase 'God save the king' was a watchword of the Royal Navy in the mid-sixteenth century, eliciting the response 'Long to reign over us'. The earliest known version of music and lyrics together in a form similar to the one used today was published in 1745 in the *Gentleman's Magazine*. This version began with the words 'God save Great George Our King' – George II's place on the British throne then being under threat from the Jacobite Rebellion led by Prince Charles Edward Stuart.

329. Who wrote the words to 'Happy Birthday to You'?

In 1893 the American sisters Patty and Mildred Hall wrote a tune that has been described as the most recognized song in the English language. It was written for children, and the original words were not 'Happy birthday to you' but 'Good morning to you'.

The 'Happy birthday' lyrics to the same tune first appeared in print in 1912, but are thought to be even older. Whether the adaptation was made by the Hall sisters, their students, or someone else entirely is unknown. The song was not copyrighted until 1935.

330. Who wrote the music for 'The Battle Hymn of the Republic'?

In the 1850s there was a popular Methodist campfire song in America known either as 'Say Brothers, Will You Meet Us' or 'Canaan's Happy Shore'. The first of those lines was sung three times followed by the words 'On Canaan's happy shore'. There followed a chorus beginning with the words 'Glory, glory, hallelujah!'

In 1861 a new version of the song appeared, with the same chorus but with the lyrics known as 'John Brown's Body' (referring to the abolitionist John Brown, who in 1859 had been hanged for his part in an anti-slavery revolt). Finally, in 1862, in the second year of the American Civil War, Julia Ward Howe published the version entitled 'The Battle Hymn of the Republic'.

The original tune is usually attributed to William Steffe, but it seems likely that he did no more than take a song that already existed in the folk tradition and arrange it for a Methodist song book, around the mid-1850s. Composition of the 'John Brown' version, with music for the verse as well as the chorus, was also claimed by Thomas Brigham Bishop, a Maine songwriter, bandleader and Union soldier, but he admitted that he was building on a traditional tune.

Others too have claimed authorship of the music, but perhaps that highly successful and prolific composer known as 'Traditional' ought to get the credit.

331. Where is the body of Glenn Miller?

On 15 December 1944 a plane took off in terrible weather from an RAF base near London to take Glenn Miller to join his band in Paris. According to the official story, neither the plane nor Miller nor his companions were ever seen again. Investigations into what happened have led to a number of theories, of which the following are only a small sample.

(i) Miller's plane was bombed or shot down by the Germans.

(ii) Miller's plane, flying low, was hit by bombs ditched by a British plane returning after a raid on Germany.

(iii) The plane crashed on the French coast, but the accident was covered up by the general who had ordered the flight to go ahead despite the wretched weather. This version was put forward in 1999 by a veteran army chaplain from the unit that claimed to have found the plane and the bodies.

(iv) Miller arrived safely in Paris, where he was killed in a brawl in a brothel. This version was given by a German journalist in 1997 who claimed he had been told the story by German intelligence experts.

See also COMPOSERS 92–100, ENGLISH LANGUAGE 176, HUMAN BEHAVIOUR 232, MURDER 322, MUSICAL INSTRUMENTS 332–4

MUSICAL INSTRUMENTS

332. Who invented the violin?

We know that the ancient Chinese and Japanese both had the idea of stringing horse hair over a box and rubbing it to make music, and that Leonardo da Vinci produced plans for a keyboard instrument that played notes by drawing a bow across strings. We can also date the appearance of the four-string violin more or less as we now know it to the early or mid-sixteenth century in Italy. We know that this four-string version developed from a three-stringed predecessor and that its design was perfected by Andrea Amati, grandfather of Nicolò Amati, the great seventeenth-century violin maker. But the identity of the first person to come up with the idea of the fourth string is lost in musical history.

333. What is responsible for the extraordinary sound quality of Stradivarius violins?

In Italy in the seventeenth and eighteenth centuries, a number of instrument makers – notably Nicolò Amati, Antonio Stradivari or Stradivarius, and various members of the Guarneri and Guadagnini families – produced cellos and violins that give a richer and more pleasing tone than anything we can make today, despite all the advantages of modern technology. So what was their secret?

Some have put it down to the climatic conditions in Italy at the time, which they contend produced the perfect wood for musical instruments. Some say it is what the makers did with the wood, such as soaking it in water or treating it for woodworm. Some say it is simply the weathering of the material over the centuries that gives the sound quality, and that the instruments probably did not sound so good when they were made. Some say it was the secret recipe of the varnish; some say it was simply the meticulous craftsmanship. Some say it was a combination of all the above.

Recent scientific tests, using X-ray spectroscopy to identify the elements in substances applied to the wood, have failed to reveal the secret. The only advance in knowledge has been a negative finding: there was nothing special at all in the varnish Stradivarius used. It was just the same combination used by other makers of that era: an application of linseed oil, followed by an oil resin containing red iron oxide and other common crimson pigments. Whatever the secret was, though, the result sounds wonderful.

334. What is responsible for the extraordinary sound quality of the banjo?

The characteristic twang of a banjo has fascinated acoustic engineers for years. The sound comes from vibration of the strings and a circular membrane, each of which set off sympathetic vibrations in the others, and this sympathetic feedback continues. Recent efforts to analyse the sound and produce computerized acoustic models of the banjo sound have made some progress, but a computer-banjo still does not quite sound like the real thing.

NUMBERS

335. Why in The *Hitchhiker's Guide to the Galaxy* did Douglas Adams choose 42 as the answer to Life, the Universe and Everything?

Adams was frequently asked this question and gave various different answers, all boiling down to basically: 'No reason at all.' 'I just wanted an ordinary, workaday number, and chose 42,' he once said. 'It's an un-frightening number. It's a number you could take home and show to your parents.'

Others have suggested that it's a pun on the song title 'Tea For Two': for tea, two. Yet the origins may have lain deep in Adams's subconscious when he was working as an extra in some management training videos made by John Cleese's Video Arts company. One such video featured Cleese as a bank teller having great difficulty adding up a column of numbers. At the end of the film, Cleese finally reaches the answer and triumphantly declares 'It's 42!' At that moment, the man walking across the back of the scene is none other than Douglas Adams.

It has also been pointed out that 9 × 6 does equal 42, just as the Earth computer claimed, as long as you are working in base 13. Adams, however, strongly denied that this had anything to do with it, saying: 'I may be a sorry case, but I don't write jokes in base 13.'

336. Why does the number 42 crop up so often in the works of Lewis Carroll?

In *Alice in Wonderland*, Lewis Carroll refers to 'Rule 42: All persons more than a mile high to leave the court', while the price on the Mad Hatter's hat is 10s 6d, which is 126 pence, or 3 × 42. The first edition of the book had 42 illustrations by John Tenniel.

In *The Hunting of the Snark* the Baker has 'forty-two boxes, all carefully packed, / With his name painted clearly on each'. The poem also refers to a Rule 42: 'No one shall speak to the Man at the Helm.'

Why was Carroll so fond of the number 42? When not writing for children, he was the Reverend Charles Lutwidge Dodgson, a Cambridge mathematics don with a keen interest in comparative religion. It has been pointed out that there is a Hindu goddess with 42 arms, and that there were originally 42 Articles of the Church of England before they were cut down to 39. Or Carroll may have identified it, just as Douglas Adams claims to have done (→ 335), as a nice un-frightening number you could take home to your parents.

337. Is the 196-algorithm true for the number 196?

Think of a number with at least two digits; reverse it; add the two numbers together. If the answer is palindromic (i.e. reads the same forwards as backwards), then stop. Otherwise, repeat the process of reversal followed by addition. To give a couple of examples:

42 + 24 = 66, which is palindromic.

87 + 78 = 165; 165 + 561 = 726; 726 + 627 = 1,353; 1,353 + 3,531 = 4,884.

It has been conjectured that whatever number you start with, you will always eventually reach a palindromic answer, but it may take some time. Starting with the number 187, for example, you do not reach the goal until the answer is 8,813,200,023,188.

The reason this is known as the 196-algorithm is that nobody knows if it is true for the number 196. If it does end up with a palindromic number, it has been shown that the number will have at least 300 million digits. But nobody has succeed in proving, for 196 or any other number, that the process will never reach the desired goal.

LEARNING, *n*. The kind of ignorance distinguishing the studious.

Ambrose Bierce (1842–1913), *The Devil's Dictionary*

338. Why is 666 the 'number of the beast'?

In the King James Bible, the Book of Revelation (13:18) has the following passage: 'Let him that hath understanding count the number of the beast: for it is the number of a man; and his number is six hundred threescore and six.'

But why 666? Various explanations have been given using arithmetic codes to translate names into numbers and identifying beastly characteristics in a variety of people from the Emperor Nero (renowned persecutor of the early Christians) to Napoleon and Kaiser Wilhelm II of Germany. But none of these theories satisfactorily explains why 666 was chosen.

It is the number you get if you add together all of the Roman number symbols – D, C, L, X, V and I – but if that is the reason, then why not include M and arrive at 1,666?

The question is further muddied by recent research suggesting a transcription error and that the real number of the beast is not 666 but 616 – but that is just as difficult to justify.

See also MATHEMATICS 278–80

OLD TESTAMENT

339. How long was the cubit mentioned in the account of the dimensions of Noah's Ark?

When the Lord told Noah to make an ark of gopher wood, He specified that 'the length of the ark shall be three hundred cubits, the breadth of it fifty cubits, and the height of it thirty cubits' (King James Bible, Genesis 6:15). Noah must have known what God was talking about, because he did not then ask whether the Lord was referring to an Egyptian royal cubit, a Roman cubit, a Persian cubit or a Sumerian cubit, all of which could refer to different lengths – a state of affairs that has led to disagreements about anything measured in cubits ever since (→ JESUS CHRIST 251).

Traditionally, a cubit was the distance between a man's elbow and the tip of his outstretched middle finger. It was also frequently defined as a fixed number of 'palms', which were the breadth of a hand, but the number of palms in a cubit might vary between six and nine, and in most cases the cubit defined in this way was longer than that given by the other definition of elbow to middle finger.

340. What were the names of the women on Noah's Ark?

The Book of Genesis lists four women on the Ark: the wives of Noah and of his sons Shem, Ham and Japheth. But we are not told the names of any of them. Other ancient scribes do suggest names for them, but no two writers agree, and around ten sets of names, often widely differing from each other, have been offered. The one thing we can be reasonably certain of, contrary to the beliefs of 12 per cent of Americans surveyed in 1997, is that Joan of Arc was not Noah's wife.

341. Did the flood that resulted in the formation of the Black Sea inspire the story of Noah?

There are two rival theories concerning the formation of the Black Sea. The older theory is that over the past 30,000 years or so, water has flowed between the Aegean and Black Seas, sometimes in one direction, sometimes the other, but gradually raising the water level in the latter. The other theory, first proposed in 1997, is of a massive flood around 5600 BC caused by the waters of the Aegean overflowing and spilling over the Bosporus at a rate about two hundred times greater than today's flow of water over Niagara Falls. The result was the formation of the Black Sea and the flooding of a vast area of land, causing at least displacement and probably disaster for all the peoples of the region.

Memories of such a cataclysmic event could, it has been suggested, have led to the story of Noah's Flood – a story that is by no means unique in Middle Eastern mythology, paralled in other sources such as the *Epic of Gilgamesh* from Mesopotamia. Scientific examination of Black Sea sediments has not yet either confirmed or refuted the hypothesis of a sudden flood, but a link to the Noah story would be even more difficult to establish.

342. Was the Exodus of the Israelites mythological or historical?

There are so many independent references to the Exodus in the Old Testament and other ancient texts that it is difficult to dismiss the story as pure myth. Those seeking a historical background for the events described in the Book of Exodus and the next three books of the Old Testament point to the late Bronze Age collapse of various civilizations in the eastern Mediterranean (→ ANCIENT HISTORY 13) and a number of examples of Egypt expelling certain peoples from its lands. The massive eruption of the volcanic island of Thera around 1600 BC has also been cited as a possible explanation of the plagues of Egypt and the crossing of the Red Sea.

Yet the account in Numbers of 600,000 men and their women and children being led across the wilderness by Moses does not tally with what we know of contemporary population figures, either in Egypt at the time of the supposed departure of the Israelites or in the promised land of Canaan at the time of their arrival. The figure must be at best an exaggeration or mistranslation. No archaeological evidence has been found in support of the Exodus story, though a number of sites have been excavated along the supposed route.

It seems reasonable, therefore, to conclude that the story of the Exodus is a blend of religious and historical elements, though we cannot be sure of the factual basis of the latter.

343. Who wrote the Old Testament Psalms?

The Book of Psalms contains 150 items, of which 73 are ascribed to David. However, these ascriptions in ancient texts are thought by some scholars to be unreliable, having been added long after the psalms were written. Another view is that the word ascribing them to David did not mean 'by David' but 'of David', meaning only that he collected together the works of various ancient psalmists. Analysis of the language used in the psalms has also been taken by some scholars to suggest that they were written over a long period, and were not all by the same hand.

The names of Asaph and the 'Sons of Korah' also appear in ascriptions, as do those of Henan, Ethan, Solomon and Moses. As with much else in the Old Testament, we do not know who the author was and do not have enough samples from known writers to be able to undertake a proper textual analysis.

See also JUDAISM 252, 254

OLM

344. How does the olm live for so long, and why doesn't it seem to age?

The olm, or proteus, is a blind wormlike amphibian around 20 to 30 cm (8 to 12 in) long that lives in underground caves in the karst region of the northern Balkans. Its eyes are undeveloped but have some sensitivity to light, while its senses of hearing and smell are good. The remarkable thing about the olm is that, according to a recent estimate, it may live for a hundred years, which is far longer than one would expect from its size. Even more remarkable is the fact that – according to recent examinations of specimens over the age of sixty – they show no physical signs of ageing. The olm can also live for up to ten years without food. Whether the absence of sunlight from its habitat contributes to its health and longevity is a matter for further research.

All you need in this life is ignorance and confidence, and then success is sure.

Mark Twain (1835–1910)

PAINTING

345. What happened to Leonardo da Vinci's lost masterpiece, *The Battle of Anghiari*?

In 1504 Leonardo da Vinci was commissioned to paint a battle scene in the Hall of the Five Hundred at the Palazzo Vecchio in Florence. Michelangelo was to paint a different battle scene on the opposite wall. It was the only time the two men worked on the same project, and it was to be Leonardo's largest work.

We have a good idea of the magnificence of the work from the artist's numerous preliminary sketches and from the copy of the central section made in the following century by the Flemish master Peter Paul Rubens, with its splendid depiction of men on horseback in the midst of the fury of battle. But Leonardo's original is lost. Sometime in the mid- to late sixteenth century, the Hall of the Five Hundred was enlarged and the painting was lost. Some think it may still be hidden behind later frescoes on the wall, and that perhaps one day it will be revealed again.

346. Did Leonardo da Vinci leave a coded message in microscopic letters in the eyes of the *Mona Lisa*?

Dan Brown's best-selling 2003 book *The Da Vinci Code* may have been fiction, but in December 2010 members of Italy's National Committee for Cultural Heritage claimed to have discovered what may be a true da Vinci code. Apparently, if you hugely enlarge images of the eyes of the *Mona Lisa*, letters and numbers can be detected within them. These can only be seen with the aid of a microscope, but the researchers say the right eye clearly shows the letters LV, which could stand for Leonardo da Vinci, while the left eye also has symbols, but they are rather blurred and it is difficult to make out what they are. The number 72, or it may be L2, is in the bridge in the background. The researchers had been prompted to examine the painting closely after coming across a fifty-year-old book in an antique shop, which indicated that such symbols might exist.

347. Who was the model for the *Mona Lisa*?

There has always been some doubt over the identity of the model in Leonardo's *Mona Lisa*, but the general opinion has always been that she was Lisa Gherardini, wife of Francesco del Giocondo, the silk merchant who commissioned the painting. This was apparently confirmed in 2005 when a note was found in the margin of a book written in 1503, identifying her as the model.

In February 2011, however, another theory emerged. Not only was it not Lisa Gherardini, it was not even a woman. Comparing the face in the painting with other faces appearing in Leonardo's oeuvre, it was suggested by the Italian art historian Silvano Vinceti that it was in fact the face of Gian Giacomo Caprotti, known as Salai, an attractive male apprentice of Leonardo's who is thought to have also been his lover. Supporters of Vinceti's theory also referred to the mysterious microscopic coded letters (→ 346) in the portrait's eyes, one of which is said to be an S, possibly standing for Salai.

Not to be outdone, in May 2011 supporters of the traditional Lisa Gherardini theory announced that they were close to identifying her burial place, and when they do so they hope to exhume the bones, find the skull, and from that recreate what she looked like.

348. Who was the subject of Albrecht Dürer's sketch entitled *Head of a Negro*, and what was Dürer's relationship to him?

Dürer's marvellous 1508 portrait known simply as *Head of a Negro* was drawn right at the beginning of the transatlantic slave trade, a time when black men were not often seen in Europe other than at ports (mostly in Spain and Portugal), where they would have been en route to America. Yet Germany, where Dürer lived and spent most of his time, had no such ports. Perhaps the rarity of the subject was what prompted Dürer's desire to draw the man, but if the subject was indeed a slave, how did he come to be sitting for Dürer, and if he wasn't a slave, then who was he?

In the art of the time, black faces were common in paintings of the Adoration of the Magi, but rare in other works, particularly portraits. In his journal, Dürer mentions meeting an African woman in Antwerp, in the house of the Portuguese trade commissioner. This woman was the subject of his portrait known as *The Negress of Brandon*, but Dürer's meeting with her took place in 1521, long after he sketched his *Head of a Negro*. The identity of the subject of the latter picture is still a complete mystery.

349. Which of Rembrandt's supposed self-portraits were actually painted by Rembrandt?

The seventeenth-century Dutch master Rembrandt van Rijn painted more self-portraits than any other major artist, but there is a debate over the precise number. The problem is that the paintings that have been classified as Rembrandt self-portraits come in three categories: the paintings Rembrandt executed himself; the forgeries; and the paintings that were executed, copied or completed by his students. The first two categories are clear enough, but the last creates a good deal of room for argument and confusion. At one time, about ninety paintings were thought to be authentic Rembrandt self-portraits; now the estimates vary between thirty and eighty.

Interestingly, New York Customs records show that 9,482 alleged Rembrandts were imported to the USA between 1800 and 1850. The worldwide figure for works confirmed as genuine Rembrandts is now down to around three hundred, though new discoveries continue to be made. In 2008 a newly authenticated Rembrandt self-portrait sold for £2.2 million ($4.5 million at the time) at auction. The auction house had thought it to be a fake and had estimated that it would fetch £1,500.

350. How did Caravaggio die?

Apart from being a great artist, Michelangelo Merisi da Caravaggio was a notorious brawler and trouble-maker. Jailed several times and exiled from Rome for killing a man in a fight, he was even expelled as official painter to the Order of the Knights of Malta as 'a foul and rotten member'.

Caravaggio died at the age of thirty-eight in 1610, supposedly of a fever, but since several attempts had previously been made on his life, one of which is said to have left him disfigured, foul play cannot be ruled out. In 2010, bones found by Italian researchers at a church in Tuscany were identified 'with 85 per cent certainty' as those of Caravaggio. Investigations suggested sunstroke, infected wounds and lead poisoning, possibly from the lead in his paints, as possible causes of death.

Ignorance more frequently begets confidence than does knowledge. It is those who know little, and not those who know much, who so positively assert that this or that problem will never be solved by science.

Charles Darwin (1809–1882)

PALAEONTOLOGY

351. What explains the 'Cambrian explosion', when a vast number of species of plants and animals appeared around 530 million years ago?

When Charles Darwin proposed his theory of evolution, one of his biggest problems came in accounting for the Cambrian explosion. Evolution ought, according to Darwin, have been a slow process, yet the fossil record seemed to indicate that over a few million years, which is a very short period in evolutionary terms, a large number of species of both animals and plants came into existence. When the Cambrian period began 570 million years ago, most organisms were single-celled creatures. Then suddenly a large range of multicellular organisms evolved, resulting in something much more like life as we know it today. There are several suggestions to account for this:

(i) The fossil record gives an incorrect picture, as something may have happened in the Precambrian era that prevented fossils from being laid down.

(ii) There's nothing wrong with the fossils. It's just that for some reason we have failed to find those in the period just before the Cambrian.

(iii) Some vital change occurred in the evolution of living organisms in the early Cambrian that gave a huge advantage to multicellular creatures, enabling them to evolve in great numbers.

(iv) Something in the climate, atmosphere or elsewhere in the environment changed the nature of what types of creature were best fitted to survive.

Since 1980 more Precambrian fossils have been unearthed that support the view that the Cambrian explosion may not have been as explosive as had been thought, but it is still difficult to explain in standard evolutionary terms.

352. How did the creatures of Madagascar get there?

Around 75 per cent of the animal species in Madagascar are found nowhere else on Earth. The fossil record has never supported a case for a slow, separate evolution on Madagascar itself, so the belief has generally been that the creatures of Madagascar must have come from the African continent long ago, either by a land bridge or carried over by ocean currents on rafts of matted vegetation.

The land-bridge theory runs into the evidence that different species arrived at different times over some 40 million years. If there was a land bridge, one would expect them to have wandered across over a relatively brief period rather than having the ancestors of shrews and hedgehogs waiting some 25 million years after lemur-like creatures had crossed. Yet the raft theory runs counter to our knowledge of ocean currents around Madagascar, knowledge that makes it unlikely that anything drifted over from Africa.

In 2010, however, a paper in *Nature* reported research supporting the view that the currents round Madagascar may have changed both in speed and direction some 23 million years ago, which was about the time new species ceased to arrive on the island. Measuring the strength of prehistoric ocean currents is a tricky business, however, and it may be that this has to remain a theory.

353. Why did mammoths become extinct?

Numerous theories have been put forward:

(i) The last mammoths froze to death in Siberia at the onset of the last ice age (Victorian theory).

(ii) Climate change and disease are the most probable causes of the mammoths' extinction (Penn State University research, 2008).

(iii) Humans hunted the mammoths to extinction (research in Madrid, 2008).

(iv) Humans hunting mammoths was not the main problem. It was human over-hunting of other animals and possibly the over-consumption of certain plants, which in turn caused major disruption to the entire ecosystem, leading to the extinction of many species (research at UCLA and Oregon State University published in 2010).

(v) Mammoth extinction was caused by the milder climate leading to a growth in trees, which increased the shade and took nutrients from the soil, both of which led to a severe decrease in the growth of the vegetation the mammoths needed to survive (research in Alaska, 2006).

(vi) Mammoth extinction was a consequence of a large change in the Earth's fauna about 13,000 years ago (conclusion of a study of fossilized mammoth dung in Wisconsin, 2009).

354. Did our *Australopithecus afarensis* ancestor known as Lucy walk on the ground or swing from the branches of trees?

Australopithecus afarensis lived between 3.9 and 2.9 million years ago, which is at least a million years before *Homo erectus*, but which of them was the first to be truly bipedal continues to be a source of debate. The discovery of *Australopithecus* footprints in 1978 demonstrated that they certainly did walk upright at least some of the time, but studies of their bones offer conflicting information.

The curvature of the fingers and toes is similar to that of modern apes and suggests that the ability to grasp branches and climb was important to them. On the other hand (or more accurately foot), the absence of an abductable (grasping) big toe – an absence shared by modern humans – suggests that their feet were used more for walking than swinging from branches. Their pelvic formation has also been cited as evidence for a bipedal gait. We do not even know if they lived most of the time in trees or on the ground.

355. Did Neanderthals have furry coats of hair?

We know, from their fossil skeletons, that Neanderthals (→ HUMAN EVOLUTION 237) were shorter than us, had bigger noses and projecting brows and not much in the way of chins. But they did have bigger brains than we do. We also know that they shared an ape-like ancestor with us, but one of the troubles with skeletons is that they have no hair. So we do not know whether Neanderthals were furry like apes or relatively hairless like modern humans.

According to DNA research, Neanderthals had a similar distribution of hair colours to our own. Or, as one somewhat dubious newspaper report of 2008 put it: 'European Neanderthals had ginger hair and freckles.' Less eye-catching reports suggested that about 1 per cent of Neanderthals had ginger hair.

When, in the course of our evolution, we lost most of our body hair, or why we did so, are unknown. One recent suggestion is that parents put their furriest babies to death through a desire to distance their species from other mammals, so passing on less hairy genes. But whether this happened (if it happened at all) before or after the Neanderthals is a matter of pure speculation.

356. Did Neanderthals have music?

DNA research has shown that Neanderthals possessed the FOXP2 gene, which is thought to play a vital part in the acquisition and use of language, so they may have been the first humans to talk to each other, or at least to grunt in an intelligible manner (→ LANGUAGE 237). There is another theory, however, that language began with music and that human speech began with something closer to birdsong. So it is possible that the Neanderthals communicated musically and, thanks to the FOXP2 gene, this evolved into language.

PANDAS

357. Why are pandas black and white?

Almost everything we said about zebras (→ MAMMALS 266) applies equally to pandas, except that pandas have no predators, so the idea that their colouring acts as a warning to other species is even less convincing. It has been suggested that the stark black-and-white markings do act as a kind of warning – a territorial warning to other pandas to keep away. It has also been suggested that the colouring enables pandas to see each other more distinctly and helps them in their search for a mate.

358. How does a female panda choose which of her cubs to bring up and which to abandon?

Pandas give birth to one, two or sometimes three cubs, but they will never care for more than one. In the event of multiple births, the mother will select one cub to look after and abandon the others, which soon die. This would make sense if she always abandoned the weaker cubs, which many have assumed is what happens, but there is no evidence to support that theory, nor are there any other findings to suggest what the mother panda is looking for when she makes her selection.

PENGUINS

359. What is the penguin population of Antarctica?

In April 2011 the US National Oceanic and Atmospheric Administration published a report linking variability in the amount of Antarctic krill with a decline in penguin population. Both krill harvesting and a decline in krill put down to global warming seemed to be depriving penguins of one of the major items in their diet. Several publications seized on this report to assert that penguin numbers had decreased by 50 per cent or more, and that the survival of baby penguins had been affected still more. Yet the original paper had taken pains to avoid making claims about the entire Antarctic penguin population, the authors making it clear that their figures related only to certain species of penguin (of which there are seventeen in all in the Antarctic) and to certain closely monitored breeding colonies.

Scientists at the British Antarctic Survey have been more wary. Addressing the question of whether penguin populations are declining, they say: 'Some species are but others are not – it depends where you look.' Adélie penguin populations have fluctuated, they say, while chinstrap penguins have decreased significantly, and gentoo numbers have risen.

On their website, the Antarctic explorers Guillaume and Jennifer Dargaud are even more explicit about how little we know:

> *Many parts of Antarctica, including some of the shores, have never been explored, so the precise number of penguins is unknown. Evaluations have been done by counting manually the*

penguin populations in some areas and extrapolating to the entire continent... But keep in mind that these are highly inaccurate statistics, most colonies not having been visited for decades, and many more having never received a visit.

The definitive Antarctic penguin census has clearly not yet been undertaken.

See also SEX 417

PHILOSOPHY

The entire subject of philosophy might be considered to consist of the things we do not know. What follows is just a small selection of the sort of unanswered questions philosophers love to think about.

360. Do we have free will?

The problem as to whether or not we have free will – a question that both philosophers and theologians have struggled with for centuries – boils down at its simplest to the compatibility or otherwise between free will and determinism. The theological version comes down to a discussion of the nature of God's omniscience. If God is omniscient, the argument goes, then He knows everything about the past, present and future, including all the decisions we, His creations, are going to make. Our actions are therefore predetermined and we don't really have free will at all.

Such an argument, however, does not require the existence of God. The scientific version is based on the idea that the universe runs according to immutable laws. Every particle in the universe is governed by these laws, whether we know them or not, so all our actions must be the result of them and thus predetermined. The biological version of the same argument would include our genetic make-up and experiences, which determine other aspects of our behaviour and the choices we think we are making.

See also INSECTS 244

361. Is free will an illusion caused by quantum uncertainty?

When the laws of quantum mechanics were discovered in the 1920s, they gave us a new way of approaching the question of free will. Suddenly, at a subatomic level, we found ourselves facing a level of indeterminacy quite foreign to the rigid picture painted by classical physics. According to the 'uncertainty principle' enunciated by the German theoretical physicist Werner Heisenberg, it was impossible to determine the position and velocity of a particle at the same time, thus making it impossible to predict its future. Furthermore, according to quantum theory, particles could be in two different places at the same time and a cat could be both dead and alive (→ QUANTUM PHYSICS 401). This idea led to the idea of parallel universes, encompassing all the possibilities of quantum uncertainty.

Advocates of free will might claim that that is what they have been saying all along. When we make a choice, we are just steering ourselves in the direction of a chosen universe. Deniers of free will could equally say that all quantum uncertainty has done is to introduce a predictable randomness into physical laws, and the existence of that randomness is what makes us think we have the ability to make choices.

362. Does a cause necessarily precede its effect?

From Aristotle in ancient Greece to the eighteenth-century Scottish philosopher David Hume and beyond, the very idea of causality seemed to insist that cause precedes effect. Hume even began his definition of of 'a cause' by describing it as 'an object precedent and contiguous to another'. Only in the mid-twentieth century did philosophers begin to ask whether this was indeed necessary.

The standard argument against cause coming later than effect is similar to the argument against travelling backwards in time: if A causes B, but B precedes A, then what happens if, after B has occurred, we do something to stop A happening?

There are two obvious ways around that dilemma. One is to say that if A is truly the cause of B, then you would find it impossible to stop A from happening once B had occurred. The other is to say that stopping A does not cause a logical dilemma. B can still happen without A. All we have done is change the cause of B.

For the past fifty years, physicists have conjectured the existence of certain particles, called tachyons, that can travel faster than light – which would mean they could travel backwards in time. And that would mean that their future (from the perspective of our time frame) could affect their past.

Not long before writing this, I overheard a conversation in which a family were discussing what the date was going to be on the following Monday. 'It's definitely the 21st,' one lady said, 'because Tuesday is Geoffrey's birthday, and that is the 22nd.' That seemed to settle the matter for me: if Monday is the 21st because Tuesday is the 22nd, then the effect has preceded the cause. Some philosophers, however, would disagree.

363. What is knowledge?

Plato said that for something to be considered 'knowledge' it must satisfy three criteria: it must be justified, true and believed. For more than two thousand years, philosophers have been arguing about whether he was right. Take, for example, the following scenario:

A celebrity is involved in a road accident and a newspaper publishes a report based on incorrect evidence that he has been killed. Someone reads this report just before jetting off on holiday to an out-of-the-way location where international news is hard to come by. A week later, the celebrity dies of his injuries. Does the holidaymaker know that he is dead? He believes it, because he read it in the paper. It is now true (though it wasn't when he read it). It is justified, because it was a reputable paper that doesn't often make such mistakes.

You could say it all hangs on what one means by 'justified', and that is basically what philosophers have been arguing about all this time.

364. Is the self identical with the body?

This is one of the 'Fourteen unanswerable questions' of Buddhist philosophy that the Buddha himself refused to answer. It is the question of whether the self is eternal and unchanging, as the eternalists believe, or whether it dies with the body, as the materialists claim. It is said in Buddhist scripture that considering such matters is a waste of time, so we shall move on to the next item without further ado.

365. Is colour a product of the mind or an inherent property of objects?

The Austrian physicist Erwin Schrödinger once wrote: 'The sensation of colour cannot be accounted for by the physicist's objective picture of light-waves.' We have such a vivid perception of the concepts we call 'red', or 'blue' or 'yellow' that it is difficult to think of them as merely ranges on a scale of wavelengths of light.

There are, of course, many things in everyday experience to which we can assign figures on a precise scale: temperature, or length, or weight, for example. Yet we can tell whether one thing is hotter than another, or bigger or heavier. Without specialized knowledge of light frequencies, however, we would be hard pressed to put colours in the right order without relying on the ROYGBIV of the rainbow. Even with that help, however, we would not claim to perceive red and violet as at opposite ends of a scale.

Some of the things we experience are subjective (philosophers use the term 'qualia' for such things), some are objective. Colours fall in between: they are our subjective experiences of an objective phenomenon. Whether they exist at all or only in our minds is a difficult question to answer.

366. When is a heap not a heap?

This is known as the Sorites paradox, from the Greek word for 'heap', and it is attributed to the Greek philosopher Eubulides of Miletus (flourished fourth century BC), whose argument went as follows:

If you remove one grain of sand from a heap of sand, it will still be a heap of sand.

A million grains of sand is a heap.

So if we remove the grains one by one, it will always still be a heap.

Eventually all the grains will be gone. So even when we have no sand at all, it is still a heap.

There is a similar argument about the hairs on the head of a non-bald man.

Bertrand Russell tried to get round this paradox by saying that all natural language is vague, but when so much of philosophy depends on the precise use of language, that does seem a bit of a cop-out.

367. Is the number of stars in our galaxy odd or even?

Even the ancient Greek philosophers said that we can never know the precise answer to that one. It used to be cited as an example of a question to which a precise answer existed, but which we can never know. Actually it's an even more difficult question than the Greeks imagined, as we would have to decide whether to include white dwarfs and black holes and other types of collapsed stars, and what to do with the stars that are thousands of light years away, so may seem to be there, but could have collapsed ages ago.

According to a recent estimate, there are about 200 billion stars in the Milky Way and about 100,000,000,000,000,000,000,000 stars in the universe. But at the end of 2010, astronomers reported that we may have got it wrong and that there could be three times as many. If we are not even sure whether the figure is 1 followed by 23 zeroes or 3 followed by 23 zeroes, the chance of ever knowing whether the precise answer is odd or even is pretty slim.

368. Do coincidences happen more often than they ought to?

The Swiss psychologist Carl Jung (1875–1961) had a belief in the inherent connectedness of things, maintaining that the world is well ordered, with events propelled to move in a coherent direction by the 'collective unconscious'. At the heart of his ideas is the concept he called 'synchronicity', which he conceived as a connecting process that did not rely on cause and effect.

Coincidence-mongers love collecting and pointing out the remarkable coincidences that happen to themselves and others, some of which seem mind-boggling in their unlikelihood.

Cold rationalists argue that we all lead very rich lives, with thousands of things happening to us every day that offer coincidence opportunities. What would really be remarkable would be if surprising coincidences did not happen from time to time. But measuring the precise frequency with which coincidences happen and calculating whether they happen more often than they ought is not an easy task.

There is a famous list of supposedly mystical coincidences linking the assassination of Abraham Lincoln and that of John F. Kennedy. Both had seven-letter surnames; both were shot from behind on a Friday; both were elected in years ending -60; both were succeeded by men named Johnson who were born in years ending -08; Lincoln was in Ford's Theatre, Kennedy in a Ford convertible; and so on and so on. But there are always plenty of coincidences to be found if you are looking for them.

See also BATS 33, THE GREEKS 217–18, INSECTS 244, REALITY 408–411

PHYSICS

369. Why is gravity so much weaker than the other fundamental forces of physics?

There are four fundamental forces that hold the universe together. We are all familiar with the electromagnetic force from our experience of magnetism and electricity; we may have heard of the strong nuclear force, which is what holds protons and neutrons so firmly together inside the atomic nucleus; and if we have delved into quantum theory, we may have run into the weak nuclear force, which acts on subatomic particles and is responsible for nuclear decay. However, even the weak nuclear force – which is only a millionth of the strength of the strong nuclear force – is a giant compared with the fourth force, the force of gravity.

If you measure the gravitational force between two particles, you need to add 30 zeroes to it to get an approximation of the weak nuclear force. This is a big problem for any physicist looking for a theory of everything.

String theory is the nearest anyone has come so far. By seeing our three-dimensional world as a slice, or 'brane', in a ten-dimensional universe, a reconciliation of the forces can be obtained. This views gravity as a force acting through all ten dimensions, of which we perceive only a feeble three-dimensional effect, while the other forces act only in our three dimensions. It all comes down to the difference between 'open strings', which are the fundamental ten-dimensional vibrating squiggles of string theory, and 'closed strings', which are the same thing, but with the ends joined together to form a loop.

Some physicists, however, think that adding another seven dimensions to the universe is inventing rather a big sledgehammer just to crack the puny nut of gravity.

370. How does gravitation work?

Newton saw gravity as a force between two objects. Einstein saw it as a consequence of the warping of the space-time continuum caused by mass. Newton believed that gravity was instantaneous. Einstein posited the existence of 'gravitational waves', moving at the speed of light. Recent experiments seem to have confirmed Einstein's view, but what his gravitational waves consist of and how they are propagated remains a mystery.

371. Is cold fusion possible?

In March 1989 Martin Fleischmann and Stanley Pons of the University of Utah reported an experiment which, if it was correct, could have changed the world. What they announced was the detection of the results of nuclear fusion in an experiment conducted at room temperature. Previously fusion had only been seen to occur at enormously high temperatures. If Pons and Fleischmann were right, their discovery could lead to an almost inexhaustible source of cheap energy.

Other scientists rushed to repeat the experiment and verify the results, but with no success. By the end of 1989 there was a heavy cloud of distrust surrounding the idea of cold fusion, and much criticism was being flung at Pons and Fleischmann for alleged experimental errors.

Yet despite this denunciation by many in the scientific community, various teams have continued to pursue the Holy Grail of cold fusion, and there continue to be sporadic reports of claims that the results of nuclear fusion have been detected in experiments at room temperature. Some still clearly believe that it may be theoretically possible.

372. Can solar energy be converted into a practical fuel?

The Sun, though clearly a massive potential source of energy, cannot be used for the vast majority of our energy needs. The trouble is the intermittent nature of solar energy as far as we on Earth are concerned, together with the problems of storing it for later use. Solar panels may be good at heating our homes during the day, and the heat may to some extent be stored in hot-water tanks, but in general the best thing that can be done with excess energy captured by solar panels is to feed it into the electricity grid.

Solar-powered cars have been developed, but at best they require an alternative source of energy to be used when it is not sunny. The most practical idea at present seems to be a battery-powered car that can be recharged with solar energy – but in practical terms the best this can offer is to extend the range of the car   beyond that offered by the battery. A car that replaces conventional fuels with solar power but still performs equally well is still a long way off.

373. Is there a limit to how fast humans can run?

A hundred years ago, the world record for the men's 100 metres was 10.6 seconds. Sprinters have slowly but surely chipped away at the record, which now stands – as of July 2011 – at 9.58 seconds (a time achieved by Usain Bolt in August 2009). Given our mass, leg length, muscle power and other physical factors, there must be a theoretical maximum speed at which humans can run, but what is it?

How much of the improvement has come through the creation of faster running tracks or better shoes and how much is due to stronger and better-trained athletes is arguable, but biostatisticians have used the 100-metre record data to attempt to predict the limits of human running speed.

Until recently, a smooth curve seemed to fit the data linking world 100-metre records to the year they were set. Gradually declining but flattening out, this resulted in a prediction of about 9.45 seconds as the ultimate humans could hope to achieve. Then along came the Jamaican sprinter Usain Bolt, who broke the world record three times in 2008 and 2009, sending the actual data plummeting below the smooth curve that had been predicted. It should have been another twenty or thirty years before such times had been reached.

The biostatisticians have gone back to their computers, but with the smoothness of the curve now badly dented, all bets are now off.

See also EINSTEIN 164–7, FUNDAMENTAL PARTICLES 192–7, MAGNETISM 261–3, QUANTUM PHYSICS 399–407

THE PLANETS

374. How did the planets form?

Astronomers believe they have a pretty good idea of how the Solar System formed. It began around 4.6 billion years ago in a huge cloud of gas and dust, most probably thrown out by the massive explosive force of a supernova somewhere else in the galaxy. The gravitational pull of this massive dust cloud attracted more material, eventually causing the cloud to collapse in on itself. Since angular momentum has to be conserved, its rotational speed increased, causing the cloud to flatten out into a disc around a very dense core.

This core eventually coalesced into the Sun, while some of the dust particles at the end of the disc were flung out into space, eventually becoming the planets – again formed by the clumping caused by the mutual gravitational pull of the particles. This explains why the planets are all to be found in more or less the same plane, which is that of the original disc.

Yet the theory still leaves much unanswered. How did the discs of both the Sun and the planets take on their present almost spherical shape? How did they lose the massive angular momentum they once had? And while it is easy to accept that dust particles clump together to form rocks in space, it is not known how these rocks stick together and build up into planets. This last question may hold the key to understanding why some stars have planetary systems and others do not.

Until these questions are answered, we cannot really say we understand how the planets were formed.

375. Is there still volcanic and tectonic activity on Venus?

Venus, which in many ways is the planet most similar to Earth, has more volcanoes than any other planet in the Solar System. An active volcano was recently identified on one of Venus's moons, yet whether any of Venus's own 1,600 or so major volcanoes is still active is unknown. In 2010, temperature readings from the surface of Venus were reported to suggest the presence of three active volcanoes, which could be continuing the process of resurfacing the planet with lava.

The underlying question is whether Venus is a totally dead clump of rock (like our Moon) or whether it is still active and in the process of formation (like our own Earth). It would help if we knew whether

Venus had active tectonic plates or whether its surface is one solid piece, but again there is considerable dispute over what lies beneath the planet's crust. As of now, the Earth is still the only planet in the Solar System that is definitely known to possess tectonic plates.

376. Why does Venus rotate in the wrong direction, and why does it turn so slowly?

Looking down on the plane of the orbits of the planets around the Sun, all but one of them are rotating in the same direction on their own axis. This is the same direction of rotation as their own orbit around the Sun. The exception is Venus, whose own rotation about its axis is retrograde – in the opposite direction to its solar orbit. It is also very slow: a day on Venus (the period of rotation on its own axis) is longer than a Venusian year (the period of its orbit around the Sun).

We do not know why this is so. If, as is generally thought likely, all the planets were formed in the same way in the same era in the development of the Solar System, they would all be expected to rotate in the same direction. Venus's rotation may have been slowed by the effect of the tides in its dense atmosphere, but a complete reversal seems to need a different explanation.

377. Are there fish on the moons of Jupiter?

The European Space Agency and NASA are currently collaborating on a mission that involves two craft in orbit around Jupiter's largest moons, Europa and Ganymede, both of which are thought to have large underground oceans. The presence of water in its liquid state always raises the possibility of the presence of life, so these two moons now top the list of extraterrestrial locations in the Solar System where life might be found. Failing that, they could offer prospects of being habitable.

In 2009 the Arizona scientist Richard Greenberg raised the possibility of fish-like creatures living in the oceans of Europa. As he says, there is no evidence at present of any life in these oceans, but the conditions are thought to be similar to those that led to life on Earth, and if this proves to be the case, he estimates there could be not just microscopic organisms but 3 million tonnes of fish swimming around.

378. What is the explanation for the Great Red Spot on Jupiter?

The biggest storm in the Solar System has been raging for at least the last 347 years on Jupiter, the system's largest planet. This storm is so huge that three Earths could fit within its area, and its winds have been measured at 614 km/h (384 mph). Visible from Earth through a telescope, it appears as a Great Red Spot, the name by which it has always been known since it was discovered in the mid-1660s. Yet nobody knows why it is red, nor what caused it. One suggestion is that its tornado-like power is constantly swirling up dust and debris from the planet's surface or from elsewhere in Jupiter's atmosphere, though what lends it its bright red colour and what gives it its energy are both unknown.

In 2000 a half-size version of the Great Red Spot was spotted on Jupiter. It started out white, then turned brown, and in 2006 became the same red as its big brother, so became affectionately known as Red Spot Jr.

379. Why is Jupiter's Great Red Spot shrinking?

Big though it is, the Great Red Spot used to be bigger. Over the course of the twentieth century, it lost half its size, and between 1996 and 2006 its area diminished by a further 15 per cent or so. The reason for the decrease is unknown, though it has been suggested that it may have something to do with the Jovian equivalent of global warming and climate change.

380. What happened to the WD5 asteroid that came close to colliding with Mars in January 2008?

In November 2007 astronomers detected an asteroid 50 metres (160 ft) across, which was estimated to have a one in twenty-five chance of colliding with Mars. By January 2008 the odds had lengthened considerably, and it eventually passed Mars at a distance equivalent to about 6.5 times the radius of the planet. Even so, that is near enough for the gravitational pull of Mars to have had a significant effect on the asteroid's path, which explains why tracking it became impossible and it was soon lost. A larger planet, such as Jupiter, might well have exerted a pull strong enough to fling the asteroid right out of the Solar System, but Mars is thought too small to have such an effect and so WD5 may well still be somewhere in the vicinity of the Earth and Mars. The relatively small size of WD5 makes it difficult to detect, and we are unlikely to see it again unless it passes close to Earth.

381. Is there, or has there ever been, life on Mars?

In 1996 analysis of a meteorite that had fallen in Antarctica about 13,000 years ago revealed traces of fossils of bacteria that suggested there may have been primitive life on Mars 4 billion years ago. An unnamed source at NASA described the finding as 'arguably the biggest discovery in the history of science'.

In recent years, space probes have produced increasing evidence that there was once water on Mars. In 2008 ice was found, showing that water is still present on the planet, and in 2009 and 2010 it was claimed that liquid water had been identified, all of which appeared to confirm that there were, and still are, the conditions to support life on Mars. In 2009 methane was discovered in the planet's thin atmosphere, raising the possibility that the gas could have been produced from a biological source.

Where this Martian water came from and how long it has been

there, however, are still puzzling scientists, and firm evidence of life, either now or in the past, is still absent.

See also THE EARTH 141–50, THE SOLAR SYSTEM 436–9

PLANKTON

382. Why are there so many different species of plankton?

From the point of view of evolution, plankton pose a bit of a problem. Under the standard evolutionary model, when two related species are competing for the same resource, the stronger should prevail and the other become extinct. The only important things plankton have to compete for are light and nutrients, yet for some reason a vast number of different species have evolved. This is all the more puzzling as their main contribution to the marine ecosystem seems to be to prop up the bottom of the food chain and be eaten by other species.

The two main theories to account for this suggest either that there are plenty of delicate environmental facts such as current and water turbulence that may have led to the emergence of different species, or that the fact that many plankton eat other plankton may be combined with the constant variation of environmental conditions to ensure that there is no point of equilibrium between the various species, and so evolution continues to drive diversification.

PLANTS

383. Where did flowers come from?

'The rapid development, as far as we can judge, of all the higher plants within recent geological times is an abominable mystery.' So wrote Charles Darwin in a letter to the botanist Joseph Dalton Hooker in 1879.

According to the fossil record as Darwin saw it, after mosses, ferns, conifers and other varieties of green plant had been around for a couple of hundred million years or so, flowering plants suddenly appeared about 130 million years ago, and very rapidly diversified into something similar to the wide range of species we know today.

Much recent research has gone into examining the DNA of a plant called *Amborella trichopoda*, which is found in the rain-forests of New Caledonia in the Southwest Pacific. With tiny greenish-yellow flowers and red fruits, it is thought to be the missing link between the gymnosperms (primitive plants without flowers) and angiosperms (flowering plants), and a direct descendant of the very first flowering plant. Its genome may help to identify how flowering plants came into existence, but their remarkably fast proliferation and diversification are likely to remain 'an abominable mystery' for some time.

384. What determines the height to which different plants grow?

Plants can be classified as either 'determinate' or 'indeterminate', according to whether they grow to a specific size then stop, or continue growing as long as they receive nutrients. In some plants, genes have been identified that come into play to stop growth when a certain criterion is reached, but there seems to be no single gene producing that function for all plants.

To add to the confusion, several plants, such as the tomato, have both determinate and indeterminate varieties. We have not yet identified the mechanism, when determinate and indeterminate varieties are crossed, that dictates which of the two growth patterns the hybrid will follow.

385. Do plants generate methane, and if so, how do they do it?

In 2006 a discovery was announced that threatened to overturn a whole range of beliefs concerning the role of plant life in the Earth's biosphere. Researchers at the Max Planck Institute in Heidelberg, Germany, were monitoring emissions from leaves and trees in laboratory conditions designed to be similar to those the plants would encounter in the open air. Much to their surprise, they detected a significant amount of methane in the emissions, where none had been expected, and the amount of methane increased when there was more sunlight and when the air was warmer.

This finding, if true, would necessitate a radical reassessment of our view of the effects of plant life on global warming. We know that plants absorb carbon dioxide, which is a greenhouse gas – which is why they are valued in the fight against global warming – but if plants also emit methane, which has a greater warming effect than carbon dioxide, they could be net contributors to the problem, rather than a method of combating it.

The German researchers estimated that plants could be responsible for 10 to 30 per cent of all methane emissions. Yet others were unconvinced. In oxygen-starved environments, such as swamps or rice paddies, methane emissions are common, yet the production of methane in oxygen-rich conditions requires considerably more energy and it was unclear where that energy would come from.

Three years later, another theory emerged: that the plants were not creating methane, but merely absorbing it from the soil and releasing it again through their leaves. For the time being, however, the question of the role of plants in methane production is still open.

386. How does the growth of trees reflect changes in climate?

Until recently, the width of tree rings had been taken as a good indication of climate. Indeed, the whole science of dendroclimatology sprang from that observation, and much of the historical data on global warming has been obtained from the study of tree rings. Research published in 2007, however, indicated a discrepancy between the climate record as indicated by reconstruction from tree-ring data and temperatures actually recorded.

Whether this anomaly is specific to the high northern latitudes where the data came from, or whether it indicates a more general problem, is unclear. As the researchers said, 'The causes, however, are not well understood and are difficult to test due to the existence of a number of co-varying environmental factors that may potentially impact recent tree growth.' In other words, it's a bit more complex than had been thought.

The greater our knowledge increases, the greater our ignorance unfolds.

John F. Kennedy (1917–63)

POPES

387. Was there ever a real Pope Joan?

Separating myth from fact in tales about the papacy in the Dark Ages can be very difficult, but there can scarcely be any tale more unlikely than that of Pope Joan. The tale concerns a learned and religious woman who disguised herself as a man, rose in the hierarchy of the Roman Catholic Church and was elected pope in 853. After two years in the papacy, she is said to have given birth while riding a horse. There are different versions of what happened to her after that: some say, for example, that she was killed by an angry mob, while others say she repented and lived out the rest of her days in obscurity.

Most early references to Pope Joan (or Joanna) date from the thirteenth century, though a disputed mention of her exists in a ninth-century document in the form of a footnote – although this may have been added later. Her existence seems to have been widely believed in throughout the Middle Ages, though most historians and religious scholars nowadays dismiss it as a myth.

388. Were medieval popes examined on a toilet-lid-like chair to establish their masculinity?

One of the consequences of the Pope Joan story, whether it was true or false, concerned a marble chair with a large, toilet-like hole in its seat. This was said to have been a part of the papal enthronement, when the pope-elect had to sit on the seat naked from the waist down, while cardinals filed past, either peering or feeling to confirm his masculinity. When satisfied, they are supposed to have incanted the line: *'Testiculos habet et bene pendentes'* in confirmation of his having passed the inspection.

This procedure is said to have been instituted in the fifteenth century in response to the widespread belief in the Pope Joan story. The truth of such a papal testing chair does not, of course, require Pope Joan to have existed, merely that people, including the cardinals, believed that she had existed. Such a chair, however, has never been produced, and in 1601 Pope Clement VIII declared the whole Pope Joan story to be a myth. Whether his *testiculos* had been pronounced *bene pendentes* before his enthronement is not recorded.

PRAYER

389. Does prayer work?

While it can hardly be denied that prayer can have a beneficial effect on the person praying, the question of whether it can influence external events has long been debated. Since the nineteenth century, many experiments have been conducted, often involving patients in hospitals, to test the efficacy of prayer. Typically one group of patients will have a team of religious people praying for their recovery, while another group will not. None of the patients or the hospital staff know who is in which group. Some experimenters have claimed their results show that prayer works; others have said it makes no difference whatsoever.

Since the gods of most religions would not play ball with experimenters trying to prove or disprove their existence by such crude means, some less obvious method of testing the efficacy of prayer is required it we are to have any hope of answering the question.

PRIME NUMBERS

390. Is there an infinite number of twin primes?

Twin primes are sets of two consecutive odd numbers that are both prime: for example, 3 and 5, 17 and 19, and 569 and 571. Such pairs have been found in which the consecutive primes each contain thousands of digits – but the question of whether there is an infinite number of such pairs is still unanswered. Euclid proved that there are infinitely many prime numbers, but twin primes are a different matter entirely.

391. Is every even number greater than two the sum of two primes?

Known as the Goldbach conjecture, this question was raised in correspondence between the German mathematician Christian Goldbach and the great Swiss mathematician and physicist Leonhard Euler in 1742. In response to one query about prime numbers from Goldbach, Euler said that it followed from every even integer being the sum of two primes, and he added 'I regard this as a completely certain theorem, although I cannot prove it.' In the centuries since, nobody else has been able to prove it either. Its truth has, however, been verified for all numbers with seventeen or fewer digits.

392. Is the Riemann hypothesis correct?

This is another of the Clay Institute prize questions (→ MATHEMATICS 281), worth a million dollars to anyone who proves it. The German mathematician Bernhard Riemann formed his hypothesis as long ago as 1859, but understanding it is difficult if you do not have a degree in analytic number theory. So let's just try to give an idea what it is about.

A century before Riemann, Leonhard Euler (him again) had been looking at the curious distribution of prime numbers – perhaps even while he was trying to solve the Goldbach conjecture (→ 391) – and in that connection invented a function (that's a sort of equation) whose value, for any given number, gave some information about the number of prime numbers smaller than any given number.

Riemann went further, extending Euler's function beyond the real numbers to the complex numbers, that is to say numbers of the form $x + iy$, where i is the imaginary number defined as being equal to the square root of -1. (I told you it was tricky.)

Riemann called his equation the zeta function, and the question that intrigued Riemann was what values of x and y made it equal to zero, for those values had an important part to play in the distribution of the primes. There were some obvious zeroes, but all the non-obvious ones seemed to be arranged on a particular line.

Riemann conjectured that all its non-trivial zeroes were on that line, and that is now known to be true for the first few billion of them. If the Riemann hypothesis is true, a whole lot of other results about prime numbers will follow, but nobody has yet succeeded in proving it, or finding a zero that is not on the magic line.

PROTEINS

393. What determines the way a polypeptide – a chain of amino acids – folds into the three-dimensional structure of a protein?

Thanks to the discovery of the DNA double helix, we are beginning to understand how a living organism assumes the form it does. The genetic code contains the recipes for amino acids, which are strung together to form polypeptides, with in turn fold up to form proteins, each of which has a specific job to do in the growth of cells.

There is one vital gap in our understanding of this process, however, which is in the folding procedure. A polypeptide is just a string of amino acids, like the beads on a necklace. This string must fold itself in the right manner into the three-dimensional shape of a protein. If it is folded in the wrong way, the resulting protein may not work and stands a good chance of being toxic. So where does the organism get the folding plans from?

The polypeptide seems to know what to do, yet it contains only the amino acids, which seem unlikely to contain all the rules needed for all their numerous applications.

394. Why are amino acids always left-handed, and sugars always right-handed?

All organic chemical compounds contain carbon atoms, which bond with other organic molecules to form the more complex molecules that constitute the basic structures of life. Sometimes the manner in which the molecules bond together allows, in theory, two different forms, which are known as left-handed and right-handed. These terms are given according to the way polarized light passes through the compound, rotating either in a right-handed or left-handed way.

It turns out that almost all our amino acids are left-handed and almost all our carbohydrates, included sugars, are right-handed. There seems no reason why we could not use right-handed amino acids or left-handed sugars, but we don't. Somewhere in the evolution of life itself, one form was ditched in favour of the other. But we do not know when this happened, or why.

395. How do proteins find their partners?

Interactions between proteins are at the heart of the life-building process, which means that a protein must have a way of finding the protein with which it is to interact, or to identify the string of the DNA code that is responsible for that protein. No one has yet discovered by what process it manages to locate the right partner in the massive stream of DNA data.

THE PYRAMIDS

396. Are the Pyramids of Giza sequenced in line with stars?

Did the Egyptians build the pyramids wherever there was room and wherever they felt like it, or was there a plan behind their placement? In 1994 Robert Bauval and Adrian Gilbert published *The Orion Theory*, which contended that the positions of the three largest pyramids at Giza were chosen to match the three stars in Orion's Belt. The pyramids are almost equally spaced, with the third just off to the side of the line of the other two. The three stars have exactly the same formation. Not only that, but the relative sizes of the pyramids almost exactly match the relative brightnesses of the stars.

Other correlations were also found between certain aspects of the pyramid constructions and various stars. There are, however, a very large number of stars in the sky, and some sort of coincidence might only be expected. Critics of the Orion theory have also suggested that the supposedly significant orientations may not have existed at the time the pyramids were built.

397. How were the pyramids of ancient Egypt built?

Every year or so, a new article appears in a newspaper or journal that claims to have 'solved the problem that has baffled archaeologists for centuries' and then proceeds to explain the latest theory of how the Egyptians built the pyramids. And they all give different answers. Building the pyramids was a phenomenal achievement. The Great Pyramid of Cheops consists of 2.3 million limestone blocks weighing roughly 7 million tonnes, and for four thousand years after it was built it remained the tallest building in the world. Even with a workforce that has been estimated to consist of 20,000 men, hacking these blocks from a quarry, dragging them to the building site (for the Egyptians did not use wheels) and lifting them to the top of this massive structure would have been an awesome task.

There is some evidence that ramps were used to lift the blocks to progressively higher levels, but the Egyptians left no written accounts of how they did it, or how the architect achieved such precision with such a huge workforce to command. So any explanation comes down in the end to an explanation of how they *might* have done it. Assuming, of course, that the pyramids were not built by extraterrestrials, as some have claimed.

398. When and why did the era of pyramid building begin and end?

The Egyptians are thought to have built pyramids as tombs for their rulers on and off for around a thousand years, from about 2700 BC until about 1700 BC, with the pyramid-building industry at its peak from 2575 to 2150 BC. In all, around 135 Egyptian pyramids are known, and they are thought by many to have been conceived as a way of ensuring the pharaoh's successful transition into the afterlife. The orientation of the hidden chambers in certain pyramids towards a dark part of the night sky has even suggested to some that they may have been conceived as some sort of launch pad into immortality.

Yet where these beliefs came from and why the pyramid building stopped are unknown. Belief in an afterlife and the rituals associated with it, as laid down in the various Egyptian *Books of the Dead*, persisted for well over 1,500 years after the end of pyramid building, so what happened to persuade the Egyptians that pyramid building was unnecessary is an interesting question.

See also SPHINX 440–44

If it rained knowledge, I'd hold out my hand; but I would not give myself trouble to go in quest of it.

Samuel Johnson (1709–84)

QUANTUM PHYSICS

399. Why do matter and energy adhere to Isaac Newton's classical mechanics in the macroscopic world, and yet, in the world of subatomic particles, they behave in a probabilistic manner?

When Newton was in charge of the universe, everything was so simple. You knew where everything was and how fast it was moving and you could work out, from his laws of motion, where everything would be in the future. Then Einstein, Heisenberg, Schrödinger, Dirac, Planck and all the other quantum physicists came along and told us first of all that we cannot know both where something is and how fast it is moving, and secondly that it is only a probabilistic wave function anyway, and the world only makes sense if we accept that it can be in two places at the same time.

In other words, the world of large objects seems to conform to the way we see things as behaving and think they ought to behave, but when we get down to the subatomic level, everything looks very different indeed. It is almost as though we are living in two radically different universes at the same time. As J.B.S. Haldane said in 1927, 'My own suspicion is that the universe is not only queerer than we suppose, but queerer than we *can* suppose.'

400. Why does Planck's constant have the value it has?

In 1900, while working on a problem concerning radiation, the German physicist Max Planck made the astonishing discovery that his theoretical predictions only matched the experimental data if one assumes that energy is not continuous, but rather comes in tiny chunks, which he called 'quanta'. This discovery led to the development of quantum physics, which changed the way we see the world. When Einstein showed the equivalence of mass and energy, it became clear that there are also discrete quanta of mass, and according to some theories there are even quanta of time.

Planck's constant, which is a measure of the size of quanta, is expressed in units of energy multiplied by time, and is equal to $6.62606896 \times 10^{-34}$ joule-second. In other words, it is equal to the energy of passing an electric current of one ampere through a resistance of one ohm for one second multiplied by a number written as a decimal point followed by 34 zeroes followed by 662606896.

The mass of all particles, including quarks and electrons, depends on Planck's constant. According to the latest theories, we can have particles with no mass, or a particle with a mass determined by Planck's constant, but nothing in between. There must be a reason for the size of Planck's chunks, which are the basic building blocks of the universe, but nobody knows what it is.

401. Is our universe the only one, or is it part of a multiverse?

The idea of stepping out of our universe into a parallel one is almost as old as science fiction itself, but it has gained a sort of respectability as a means of explaining some of the weirder parts of quantum theory. Take Schrödinger's cat, for example, a creature dreamt up in a 1935 thought experiment by the Austrian physicist Erwin Schrödinger: this unfortunate animal is both dead and alive at the same time until we take a peek and see which it is. There's no problem if you think of the experiment occurring in two universes, with a dead cat in one and a live one in the other, with no way of telling which you are in until you look at the cat.

String theory (→ PHYSICS 369) also predicts multiple universes. Well, actually there's only one, but it has ten dimensions, and our piddling little three-dimensional universe is just a slice of the real thing – and there are plenty more like it. Even the Big Bang, according to string theory, was nothing more than a collision between two slices of the universe. The idea may also help to explain dark energy (→ COSMOLOGY 108) if we see it as emanating from a force exerted on our universe by another one.

Whether we can ever envisage an experiment that would tell us if other universes exist, however, is another matter entirely.

402. If there are other universes, do they have the same physical laws as ours?

According to an idea called the 'anthropic principle', the laws of physics as we see them are only that way because we are here to see them. A number of laws and physical constants appear to be perfectly tuned to hold the universe together and create the conditions in which we have evolved. Without the inverse square law of gravity, for example, everything would fall apart; without the laws of conservation of energy and momentum, the stars and planets would not remain in their orbits. Even a small change in some of the physical constants would cause everything to disintegrate.

One way of looking at this is to say that God made it this way; a slightly more irreligious version would maintain that God had no choice in the matter if He wanted His universe to work. In contrast, adherents of the anthropic principle hold that things could have been different, but they would have been so different that we would not have been here to see them. Our view of the universe and the laws that govern it is heavily biased by our own perspective.

Whether one sees the ten-dimensional world of string theory as the one true universe of which we are but a slice, or a collection of all possible universes, or one of a number of multidimensional universes, or just a mathematical model that may explain some of the more puzzling aspects of our own universe is, until we learn a good deal more, just a matter of choice.

403. Why is there so little antimatter?

According to our current view of the universe, it all began in a Big Bang of energy, which transformed, according to Einstein's $E = mc^2$ equation, into matter. But both theory and experiment show that when energy transforms into matter, an equal amount of antimatter is created. Antimatter is just the same as matter, except in an antimatter atom, the negatively charged nucleus is orbited by positively charged positrons, while in matter, the atoms have a positively charged nucleus orbited by negatively charged electrons. Thus when matter meets antimatter, they annihilate each other and turn back into energy.

Just after the very beginning of the universe there must have been equal amounts of matter and antimatter, yet finding even the tiniest particle of antimatter is extraordinarily difficult. So where did it all go?

404. What happened in the first billionth of a second after the Big Bang?

The Big Bang happened at 1830 GMT on 28 March 1949. That at least was the precise moment when the astronomer Fred Hoyle, speaking on BBC radio, coined the term 'Big Bang'. Hoyle did not actually believe in the Big Bang theory, but was explaining the difference between the 'steady state' theory to which he subscribed and the alternative idea of a sudden creation.

In the time since 1949, more and more evidence has accumulated in favour of the Big Bang. In particular, measurements of the expansion of the universe and the speed at which galaxies are moving away from each other support the view that they all came from the same spot 13.7 billion years ago.

Theories about the creation of everything from fundamental particles to galaxies also tie in with the Big Bang theory. In fact, we can trace everything back to the tiniest fraction of a second after the Big Bang, and everything fits – but the first billionth of a second remains a Big Problem. The massive burst of energy must have created matter and antimatter in equal quantities (→ 403, 406), but something then happened to create an asymmetry, favouring the matter. The remaining antimatter was then annihilated by collisions with matter, creating the stream of energy known as the cosmic microwave background and leaving the surplus matter to form our universe. But what was responsible for the antimatter disappearing trick in that first billionth of a second is anyone's guess.

405. How do entangled particles communicate?

The idea of entangled particles, which has been confirmed by experiments, is one of the weirdest of all weird things in the weird world of quantum theory. Under certain circumstances, a pair of particles may be created that mirror each other's properties. The most common example is a pair of photons, which have a property known as 'spin' that may be either 'up' or 'down'. But if one of the pair is 'up' the other must be 'down', and this difference is maintained however far apart the particles are.

Actually, it's more complicated than that, as spin is one of those being-in-two-places-at-the-same-time sort of quantum properties that remain undecided until the particle is observed. So both the particles are in both 'up' and 'down' states until we measure them. And at that moment, as soon as the state of one is determined, so is the other.

It is almost as though the particles are communicating with each other, and experiments have shown that it happens instantaneously. Yet if information is being transmitted between the particles, it ought not to be able to travel faster than the speed of light. But it does. The optimistic view is that this may hold the key to instant intergalactic communications. The realistic view is that we don't have a clue what's going on.

406. Does antimatter come from other galaxies?

On its last trip, in May 2011, the US Space Shuttle *Endeavour* took a massive piece of equipment called the Alpha Magnetic Spectrometer to the International Space Station, where it is to be used for a variety of experiments, including the detection of dark matter and antimatter particles. In particular, it will be looking for an anti-helium nucleus which, if found, would support the theory that there may be galaxies made entirely of antimatter elsewhere in the universe. It would also lend support to the idea that antimatter particles – which have, on rare occasions, been detected on Earth – may have extragalactic origins.

407. How can we explain the wave-particle duality of light?

In classical physics, matter comes in particles while energy comes in waves. In the seventeenth century there was an intense debate over whether light was transmitted in waves or by tiny particles called corpuscles. Some aspects of light could only be explained by one, but other aspects favoured the other. In the twentieth century Einstein muddied the picture by showing that matter and energy could be transformed into each other, but this supported the highly convenient use of either particles or waves to explain physical phenomena, whichever did the job better. Especially in the subatomic world of quantum mechanics, the use of both waves and particles became not only common practice, but seemed to be essential.

One view is that matter and energy, especially light, can behave either as waves or particles (which we now call photons), but not both at the same time. Another view is that all this is deeply paradoxical and makes no sense. But it seems to work, and such paradoxes are not unusual in quantum theory.

See also FUNDAMENTAL PARTICLES 192–7, PHILOSOPHY 361, PHYSICS 369–71

REALITY

408. Is there such a thing as reality?

Plato took the view that all nouns refer to things that exist and are therefore real. Thus, because we have the abstract nouns Beauty and Good, for example, there must be real yet immaterial 'Forms' or 'Ideas' that embody these qualities in an absolute way, and of which examples of beauty or goodness that we experience in the world of the senses are mere shadows. For most thinkers, that is much too simplistic a view of reality. A noun, they would say, is merely our shorthand to describe our subjective experience or perception of something. The eighteenth-century Anglo-Irish philosopher Bishop George Berkeley took this further with the view that nothing exists unless it is perceived – and even Einstein questioned whether we can be sure the Moon is there if we cannot see it.

One of the fundamental questions of ontology (the study of the nature of being) is whether something can be said to have an existence independent of the way it is perceived. Quantum physics (→ 399–407) shows that the world is very different from the way we see and experience it, so where does that leave reality?

409. If so, how can we tell what is real?

In Daoist tradition there appears the following tale: 'Lao Tzu fell asleep and dreamt he was a butterfly. Upon wakening he asked, Am I a man who has just been dreaming he was a butterfly? Or a sleeping butterfly now dreaming he is a man?'

Looking on the Internet, we can find out how to tell diamonds from cubic zirconia, how to distinguish between real pearls and fake pearls, how to tell real fur from fake fur, and even how to tell real breasts from implants. But telling reality from illusion is a much bigger problem.

410. Are we living inside a computer simulation, as in the film *The Matrix*?

There is absolutely no way of knowing whether or not we are no more than virtual figments in a computer simulation, though in the absence of any suggestion of who is running the simulation, we tend to disregard this possibility. Perhaps the best argument against *The Matrix* hypothesis is the weirdness of the world and its physical laws. Surely nobody, even the most perverse of aliens, would have written a program incorporating such a bizarre universe. Unless, of course, it's a brilliant double-bluff.

411. Are the three spatial dimensions we perceive part of a ten- or eleven-dimensional reality?

We have mentioned before the mysterious ten-dimensional world of string theory (→ FUNDAMENTAL PARTICLES 192, PHYSICS 369, QUANTUM PHYSICS 401-2), but in fact the mathematics of the theory works equally well in ten or eleven dimensions. They are simply different formulations of the same theory. So if the maths works in either case, does it make any sense to talk of a ten-dimensional universe or an eleven-dimensional universe as 'reality', especially from our own very limited three-dimensional viewpoint?

ROME

412. What was the population of Rome at the time of the emperor Augustus?

It is frequently maintained that the population of Rome passed the million mark at the time of the birth of Christ, when Augustus was emperor, but there is precious little precise information on which this figure can be based.

According to Augustus' own census returns, there were 4,233,000 'citizens' in the Roman empire in 8 BC and 4,937,000 in AD 14, but that was in the whole empire, and we do not know who was counted as a 'citizen' anyway. The figure is thought to exclude slaves, women and children, and probably only included men available for military service. On the basis of these figures, an estimate of around 56 million has been given for the total population of the Roman empire.

As for the population of Rome itself, estimates have been made from the total size of the empire, from the area of Rome together with an informed guess at the population density, and from the figures for grain imports and consumption. These result in anything between 500,000 and 1,300,000.

For in much wisdom is much grief: and he that increaseth knowledge increaseth sorrow.

Ecclesiastes 1:18

SEX

413. How did sexual reproduction evolve?

The earliest organisms reproduced simply by splitting in two, so creating an identical copy of themselves. It's simple; it's effective; it allows for some introduction of genetic diversity through mutations; and it doesn't require the tiresome business of finding a member of the opposite sex, luring it into mating, and then successfully completing the mating process. Somewhere along the evolutionary line, sexual reproduction came into existence, but how it established itself successfully against the easier version is a bit of a puzzle.

Presumably, it has something to do with the vastly increased speed of genetic change when a mother and father both contribute half the genes to an offspring, which may have the effect of eliminating unsuccessful genes more quickly and enabling faster adaptation to changing conditions and newly emerging diseases. Yet nobody has convincingly demonstrated that such potential advantages outweigh the problems that sexual reproduction involves.

414. Is there a connection between sexual reproduction and death?

Very simple creatures can live forever. An amoeba, for example, reproduces by splitting in two; each half then becomes another amoeba, which can also split in two; and so the process continues – at least until something comes along and eats it. Even organisms that have a slightly more complex way of reproducing, which may involve merging with others before splitting into two, do not die. There is a theory that death began with the evolution of sexual reproduction.

Several creatures, from salmon to spiders, die almost immediately after contributing their sperm (in the case of males), or laying their eggs (in the case of females). The English evolutionary biologist Richard Dawkins famously wrote of the 'selfish gene', but the whole business may be even more selfish than he suggested. Once reproduction has been assured, the gene has little use for the body so is content to let it die. Perhaps we could live forever if we didn't have sex. A long life, if rather dull.

415. Why do African jumping spiders prefer to mate with spiders that have just eaten a mosquito that has gorged itself on mammalian blood?

Vampire spiders may sound like the stuff of rather lurid horror films, but in 2005 a species of spider was found in East Africa that seems to like the taste of blood. The *Evarcha culicivora* spider, however, is not likely to jump on you and sink its fangs into your neck. Not yet, anyway. Its taste for blood has so far been identified only in its preference for eating mosquitoes that have just gorged themselves on mammalian blood. Even more interesting is the fact that males of the species prefer to mate with females who have just eaten such a mosquito.

Experiments in 2009 confirmed this by wafting the smell of female spiders over males and seeing which ones seemed most attractive. The results showed that the male spiders definitely went for the females who had wolfed down the blood-gorged mosquitoes. Curiously, it didn't seem to be the smell of the blood that was attracting them: they were not at all interested in other males that had just eaten a blood-filled mosquito.

One theory is that a chemical reaction in the female's body while processing the blood may produce a chemical that attracts the male; another idea is that the smell of blood on a female may be taken as a sign of strength and breeding fitness. Further experiments are planned to test these hypotheses.

416. How and why is the sex of turtles determined by the temperature at which the eggs incubate?

Turtles lay eggs in holes they excavate on sandy beaches, and then leave them there to incubate. The eggs are very sensitive to temperature, and those that are incubated below 23°C or above 33°C are unlikely to hatch at all. When the temperature is between about 28 and 30°C, they hatch with a roughly equal rate of males and females, but when the temperature is between 23 and 28°C, the hatchlings are usually male, while for temperatures between 30 and 33°C, they are usually female.

Some birds and various other reptiles also display what is known as temperature-dependent sex determination, but neither the mechanism by which it happens nor its evolutionary significance is understood. It has been suggested that it allows the mother turtle to choose the sex of her young, but there is absolutely no evidence that female turtles do so.

417. Why are so many female albatrosses lesbian?

Albatrosses at a colony in Hawaii, according to a study published in 2008, are a model of successful lesbianism. According to the report, 31 per cent (39 out of 125) of the Laysan albatross pairs at the Kaena Point breeding colony in Oahu were female–female. Not only that, but sixteen chicks were hatched by the lesbian pairs, ten of which had been fathered by males who already had female mates of their own. 'Fathers of chicks in female–female pairs were located at varying distances from the nest and were not simply the nearest neighbours', the report continued, indicating some degree of selectiveness among the females in the matter of choosing the fathers of their offspring. Despite this, albatrosses are well known for their monogamous tendencies, and this also applies to lesbian couples, who tend to stay together for life.

Following the publication of the Hawaiian study, albatross

colonies elsewhere were examined for lesbianism, leading to a report of a similar pair of female albatrosses raising a chick together in New Zealand. The manager of the Taiaroa breeding colony was quoted as saying: 'It's an unusual situation because we've had a triangle with one male and two females for the past couple of years, and obviously that hasn't been terribly conducive to getting on with a breeding programme. This year the male left the trio, but obviously not before he had mated with one of the females. The same report also mentioned a nearby pair of male penguins who were incubating an egg together.

418. Why does the single-celled *Tetrahymena thermophila* have seven sexes?

The unicellular *Tetrahymena thermophila* is a simple creature. It is covered with a furry coat whose hairs, called cilia, wave back and forth, propelling it through the water in which it lives. And it has seven sexes, which researchers have very sensibly called I, II, III, IV, V, VI and VII.

Each of these sexes can mate with any of the other six, but proportions of the seven sexes are not equal in any large community. Naturally, this gives an individual greater choice in finding a mate, with roughly six out of seven of the community available compared with about 50 per cent in the more usual two-sex species. But there are some rather complex rules about what sex offspring can come from different pairings. Quite how such a simple creature evolved in such a sexually diverse way, or what the advantages of its doing so might be, have not yet been established.

419. Why do males ejaculate such a vast number of spermatozoa?

The average ejaculation of the human male contains around 180 million sperm, though the range has been estimated to be between

40 million and 1.2 billion. Yet it only takes one sperm to fertilize a female egg, so why are there so many – not just in humans, but in all mammals? The usual reason given is that it increases the chance of fertilization, but such a vast number looks like overkill. Fewer fitter and stronger spermatozoa would surely improve prospects more effectively.

Another reason that has been offered relates to sperm competition. The argument goes that if several males mate with the same female, each one wants to improve his chances of fertilizing her egg, and the best way to do this is by having the largest number of sperm in the competition. Yet fertilization does not take very long, and it would surely be a better strategy to be the first to mate with the female and then keep her away from other males for a short period afterwards. As it is, a vast number of sperm seem to be giving their lives in vain.

420. What (if it exists at all) is the biological root of sexual orientation?

Periodically, an announcement is made of the discovery of a 'gay gene', but this always seems to be followed by another study casting doubt on the earlier finding. Studies of twins seem to confirm the existence of a genetic component in homosexuality, while a large number of studies have claimed to identify biological differences between gay and straight people.

In no particular order, all the following have been associated with gay/straight differences: the size of certain parts of the brain; reactions to sex pheromones; responses to certain neurochemicals; ratio of limb-length to overall height; the functioning of the inner ear; the direction of hair whorls; left-handedness; finger-length ratios; ridge-density of fingerprints... and many others. Whether there is a single factor underlying all of these has yet to be discovered.

421. Was Queen Elizabeth I a virgin?

They named the state of Virginia after her, so she must have been, mustn't she? Well, not exactly. Her statesmanlike refusal to marry a foreign prince, to preserve England's independence, and her refusal to marry an English nobleman, for fear that her choice would cause discord, may have suggested a determination to preserve her chastity, but marriage and virginity have never been the same thing. There are, in fact, two prime candidates among those who may have been Elizabeth's lovers. Curiously, and perhaps coincidentally, she bestowed the title of Master of the Horse on both of them.

First there was Robert Dudley, Earl of Leicester, the childhood friend whom Elizabeth made Master of the Horse as soon as she became queen. She is said to have behaved in an extremely flirtatious manner towards him, but distanced herself from him when his wife died in suspicious circumstances (→ ENGLISH HISTORY 173). She later recommended that he marry Mary Queen of Scots, but the latter turned him down on the grounds that she would not take Elizabeth's discarded lover.

Then there was Robert Devereux, Earl of Essex, who became Master of the Horse in 1587. He also became a favourite of the queen, who is said to have loved his eloquence and keen mind. But he rather botched his job as Lord Lieutenant of Ireland, came home in disgrace, led a rebellion against the queen, and was executed.

422. Was Queen Christina of Sweden a man?

Queen Christina (1626–89) was an extraordinary woman – if she was a woman. When she was born, she was mistaken for a boy, apparently on the grounds of being very hairy and screaming with a hoarse voice. Her father, the king, who had wanted a son, gave orders that she should be brought up as a prince, and when she took the royal oath, it was as king, not queen.

Christina never married, never produced an heir to the throne, and when she abdicated in 1654 in favour of her cousin, she headed for the Danish border, changed into men's clothing and rode through Denmark as a man. Her masculine voice and general appearance were often remarked upon, and she had always shown a preference for men's clothing. In 1965 an examination of her remains reached no conclusion about her sex other than a suggestion that she may have been intersex, which is neither one thing nor the other, but by definition involves an atypical configuration of the elements that usually separate male from female.

See also GIRAFFES 216, SQUIRRELS 453, WORMS 481

SHAKESPEARE

423. What did William Shakespeare do between leaving school aged fourteen and marrying Anne Hathaway at the age of eighteen?

Almost nothing is known for certain about Shakespeare's early life, but it is generally accepted that he went to the King's New School in Stratford-upon-Avon from the ages of seven to fourteen. At the age of eighteen he married the twenty-six-year-old Anne Hathaway (some say in haste, as the marriage banns were read only once instead of the usual three times). But there is not a shred of evidence for anything he did in his formative years between the ages of fourteen and eighteen. Shakespeare's plays and poetry display not only an impressively wide vocabulary and an unequalled mastery of the English language, but also an accumulation of wisdom and knowledge in so many spheres that it is difficult to reconcile with the view that his formal education stopped at the age of fourteen. Yet no records survive to tell the tale of Shakespeare's teenage years.

424. What did Shakespeare do between 1585 and 1592?

Six months after the wedding, Anne gave birth to a daughter, Susanna. Two years later, twin children appeared: Hamnet, Shakespeare's son, and Judith, his daughter. They were baptized on 2 February 1585. After that, the trail of Shakespeare's life goes dead again until he is mentioned in accounts of the London theatre scene in 1592.

Shakespeare's 'lost years' from 1585 to 1592 have been the subject of intense speculation, with suggestions that he worked as a schoolmaster, or a theatre manager, or fled Stratford to escape prosecution for deer poaching. Such evidence as there is for any of these tends to be based on tales of people with a similar name to Shakespeare.

Shakespeare's son Hamnet died of unknown causes at the age of eleven in 1596, but we do not know whether his short life was always plagued by illness. If it was, perhaps Shakespeare spent much of this period looking after a sick child.

425. Who was 'Mr W.H.', to whom Shakespeare dedicated his sonnets?

'To the onlie begetter of these insuing sonnets Mr W.H. all happiness and that eternitie promised by our ever living poet.' So begins the dedication in the first edition of Shakespeare's sonnets, published in 1609. A few more words follow, ending with the initials T.T.

Those initials are assumed to stand for the book's publisher, Thomas Thorpe, but the identity of Mr W.H. has remained a mystery, though there has been no shortage of speculative suggestions. One of the candidates is William Herbert, Earl of Pembroke, on the grounds that he was also the dedicatee of the First Folio of Shakespeare's plays – but an earl would hardly have been addressed as plain 'Mr'. Other suggestions include William Hart (Shakespeare's nephew, who was only nine years old at the time); William Haughton, a dramatist; an actor named Willie Hughes (for whom no evidence exists that he even existed); Henry Wriothesley, Earl of Southampton (who was not only not a Mr but had his initials the wrong way round). Bertrand Russell suggested that W.H. might be a misprint for W.S. and refer to Shakespeare himself.

Frankly, all we know is that the dedicatee's initials were probably W.H. and that he was probably not a titled person. And to add to the problem, we do not even know what the word 'begetter' is meant to mean in this context. It could mean the author, but it could mean someone who inspired the sonnets, or someone who encouraged the writer to produce them.

426. Who was Shakespeare's Dark Lady?

Shakespeare's sonnets 127–152 concern a mysterious 'Dark Lady', who we learn has black hair and dusky skin. But who she was, or even whether she was a real person or an idealized figment of his imagination, is not known. Among the suggestions are Mary Fitton, who was a Maid of Honour to Queen Elizabeth and a mistress of William Herbert, Earl of Pembroke (who, you remember, may have been Mr W.H.), whose child she bore. Another candidate is Emilia Lanier, who was another mistress of a nobleman who had connections with Shakespeare, and who fitted the black-hair-and-dusky-complexion description perfectly.

Whoever the Dark Lady was, however, she would surely not have liked William Wordsworth's description of the sonnets written to her, which he thought were 'abominably harsh, obscure and worthless'.

427. Did Shakespeare really die on his birthday?

It has become conventional to celebrate the date of Shakespeare's birth as 23 April 1564 and the date of his death as 23 April 1616. There is a nice symmetry about it, especially as 23 April is also St George's Day. We know that was the day he died, but – as with most people at this period – Shakespeare's date of birth was not recorded, only the date of his baptism, which took place on 26 April. Being baptized three days after birth would not have been unusual at the time, so it is quite possible that he was born on 23 April, but there is no real evidence of it.

SLEEP

428. Why do we need to sleep?

All animals sleep. Even nematode worms sleep. Giraffes only sleep two hours a day, but they still sleep. Sleep must be very important, but we do not know why. One theory is that sleeping allows the body to repair damaged cells; another idea is that sleep allows us to replenish the molecules that are needed to transfer energy between our cells; a third idea is that sleeping is when our brains have a chance to clean out the junk that has accumulated in them during the day; and a fourth suggestion is that sleeping is necessary for our brains to be able to run through the information they have received during the day and file it away for later use.

The one thing we do know, however, is that sleep deprivation can have drastic consequences, including intense paranoia and even death. Surprisingly, we may have a good picture of what these consequences may lead to, but it still does not help us discover what sleep actually does.

429. What is the cause of the 'hypnic jerks' that sometimes occur as we are falling asleep?

Hypnagogia is the name for the state between wakefulness and sleep just as you are starting to nod off, and that is when you may experience one of those infuriating hypnic jerks, when your body jumps as though startled and may jolt you back into wakefulness. It is an involuntary muscle contraction called a myoclonic twitch. One theory is that changes in temperature, breathing and muscle relaxation may all cause the brain to think you are falling, so it sends a signal to your limbs to wake up. Another suggestion is that it is purely a muscle spasm and the brain has nothing to do with it. It's very irritating, whatever it is.

430. Why do koalas sleep so much?

Koalas sleep between eighteen and twenty-two hours a day. Even sloths are not that slothful. The reason seems to have something to do with their very slow metabolic rate and their diet. A koala eats mainly eucalyptus leaves, which are low in nutrition and very high in fibre, making them difficult to digest. A slow metabolism and a good deal of sleep may help provide the time for their digestive systems to extract all the energy they can from the leaves. But why they cannot do that while staying awake is puzzling. Cows and sheep have diets that are low in nutrition, but make up for it by spending a great deal of time eating. Koalas adopt the alternative strategy of sleeping. But falling asleep most of the time cannot be good for one's chances of survival.

431. Why do we dream?

Current theories of the role played by dreaming include the following:

(i) When we dream, we are rehearsing our behaviour in threatening situations; that's why so many dreams are horror stories.

(ii) Dreams are what happens when our brains are throwing out the rubbish from the day's experiences. That's why dreams are so disorganized.

(iii) Dreams are when our brains reset their connections in a more useful way.

(iv) Dreams are our own form of psychotherapy. As Sigmund Freud put it, dreams are 'disguised fulfilments of repressed wishes'.

(v) Dreams are just the result of our brains freewheeling after a hard day and mean nothing at all.

432. What is the connection between REM sleep and dreaming?

In the early 1960s a number of experiments were conducted that appeared to show that dreaming was just as important to us as sleep itself. The experiments were based on the belief that dreaming coincided with periods of so-called REM (rapid eye movement) sleep, when a fluttering of the eyelids indicates a darting about of the eyes while asleep. These early experiments concluded that waking people up when REM sleep began made them highly irritable – in fact noticeably more irritable than other people who were woken up just as often, but not when their eyes were flickering.

It was subsequently found that the association between dreaming and REM sleep was not so clear. People dreamt without REM and they showed REM behaviour when not dreaming. So dreaming may not be as vital as had been thought.

433. How does sleeping relate to learning?

In the 1950s some successful marketing was done around the theme of sleep-learning. You put on a tape with a lesson on it, fell asleep, and in the morning you would have learned at least some of the information on the tape. Detailed scientific investigations, however, thoroughly discredited the idea. You don't learn anything if you are really asleep. On the other hand, more recent experiments have suggested that sleep can consolidate knowledge.

One typical test involved teaching people something in the morning, then testing them twelve hours later. Their results were compared with a group who had learned the same thing at night, then had a good night's sleep and were tested in the morning, again twelve hours after the teaching. The results of the second group were 20 per cent better than the first group, though one could argue that it was merely being awake and doing something else that caused forgetfulness in the underperforming group. Similar improvements in performance, however, have also been found in animals that are taught something then allowed to sleep.

SMOKING

434. By what mechanism does smoking cause lung cancer?

We rely on our genes to ensure the smooth running of our body's normal growth processes. Any form of cancer involves a breakdown of those processes. In the case of smoking, our understanding is that the carcinogens in tobacco smoke convert to forms that react with DNA, and the altered DNA leads to the genetic changes that exist in human lung cancer. A great deal of work is now being carried out to fill in the details of this process.

435. Why do some long-term smokers develop lung cancer while others don't?

In the UK around 80 per cent of lung cancer cases are associated with smoking, and about 15 per cent of smokers develop the disease. So what prevents the other 85 per cent of smokers from developing cancer (although they may very well succumb to one of the many other fatal conditions caused by smoking)? And what causes the lung cancer in the 20 per cent of sufferers who do not smoke?

A good deal of evidence suggests that lung cancer runs in families, so the hunt for a genetic component of the disease has been intense. A number of candidates for a 'lung-cancer-susceptibility gene' have been discovered, while much attention has also focused on breakdown of the normal gene-repair mechanism that ought to prevent the disease developing. The extent to which a lung-cancer-susceptibility gene has a role in the development of the disease, or a lung-cancer-prevention gene protects against it, are questions that have yet to be answered.

The difference between what the
most and the least learned people know
is inexpressibly trivial in relation
to that which is unknown.

Albert Einstein (1879–1955)

THE SOLAR SYSTEM

436. How big is the Solar System?

The lack of clearly marked boundaries between one star's territory and another's makes this a difficult question to answer, but there are several other factors that make it even harder. The first problem is a matter of definition: where ought one to say that the Solar System ends? One answer is to say it ends where the Sun's gravitational effect is equalled by the gravitational pull of the rest of the galaxy. We cannot be sure where this is, but it is estimated to be about two light years away. The trouble is, we do not really know what is out there, two light years away, so we cannot calculate the overall gravitational pull.

The other definition of the edge of the Solar System is that it is where the solar wind ceases to have an effect. The solar wind is the stream of charged particles ejected from the Sun's upper atmosphere. It flows out across the region of space known as the heliosphere until it reaches an area known as the heliopause, where it meets equally strong solar winds from other stars. If we use that as our definition of the size of the Solar System, then estimates of its diameter vary between about 15 and 40 light hours, rather than two light years.

Even that is open to doubt, as measurements of cosmic wind by the *Voyager 1* spacecraft, which is now close to the heliopause, are significantly different from the predicted values, suggesting that we may have got it all wrong.

437. Why does space suddenly become very empty beyond the Kuiper Belt?

Travelling towards the furthest extremities of the Solar System, the last large object we know about is the planet (now reclassified as a dwarf planet) Pluto, which is one of the tens of thousands of bodies making up the Kuiper Belt. This region of the Solar System extends between about 30 and 55 AU from the Sun (an AU, or Astronomical Unit, is the average distance between the Sun and the Earth). After that, there is nothing – which is hard to explain. After all, the region beyond the Kuiper Belt is still well within the gravitational pull of the Sun, so one would expect some sort of space debris to have been captured in solar orbit.

One theory is that right at the edge of the Solar System, far beyond the Kuiper Belt, there is something called the Oort Cloud, composed mainly of frozen water, methane and ammonia. So far, the existence of the Oort Cloud is purely theoretical, with strong reasons to believe it is out there, but no definite observations or evidence to confirm it. Another theory is that we have missed one large object, sometimes known as Planet X, which may, through its own gravitational effect, have swept up all the material in the region beyond the Kuiper Belt.

Some have even equated Planet X with the planet Nibiru of ancient Sumerian mythology, which is destined to collide with Earth and bring about the end of the world. Naturally, that has been linked to the end of the world predicted for 2012 (→ INTRODUCTION) but astronomers consider that highly unlikely.

438. Why have the *Pioneer 10* and *Pioneer 11* space probes gone slightly off track?

The deep-space probes *Pioneer 10* and *Pioneer 11* were launched in 1972 and 1973 respectively, and are now both heading in opposite directions towards the furthest reaches of the Solar System. Their positions have been tracked throughout their journeys, and at first their courses went according to prediction, but since the early 1980s they seem to have been going off track in both direction and speed.

Various explanations have been offered, such as observational error, effect of the solar wind, the influence of dark matter, or gravitational effects from some large object we do not know about. It has even been suggested that there could be something wrong with our laws of physics. None of these seem entirely satisfactory solutions.

He must be very ignorant for he answers every question he is asked.

Voltaire (1694–1778)

439. Has life on Earth spread to other places in the Solar System?

Just as the theory of exogenesis proposes that life on Earth originated elsewhere in the Universe and came here through space (→ EARTH 150), so the theory of panspermia proposes that life may exist elsewhere in the universe, having spread there from Earth. In 2000 NASA scientists announced their conclusion that single-celled organisms from Mars could theoretically have seeded the Earth with life, or vice versa. However, their calculations suggest that the transfer of microbes from one solar system to another is unlikely, but they did not rule it out. It would be a little disappointing, however, to find extraterrestrial life, only to discover that it had come from here in the first place.

See also THE MOON 311–2, THE PLANETS 374–81, THE SUN 454–5

THE SPHINX

440. By what name was the Great Sphinx known to the people who built it?

For all their hieroglyphics, the ancient Egyptians were very remiss when it came to leaving behind proper documentation of some of their finest achievements. The Great Sphinx of Giza is the largest monolith statue (carved from a single piece of rock) in the world, yet we do not even know what it was called.

The name 'sphinx' was given to it by the ancient Greeks around 500 BC, in reference to a Greek mythological beast of that name with a lion's body, a woman's head, and wings – though the Egyptian statue has no wings and the sex of its head is arguable. About a thousand years earlier the Egyptians had referred to it as Hor-em-akhet ('Horus of the horizon', Horus being one of the main gods of ancient Egypt), but the statue was originally made more than a thousand years earlier still, but during that millennium no record of any name has been found.

441. How old is the Sphinx?

The standard theory is that the Sphinx was carved around 2500 BC in the reign of the Pharaoh Khafra, yet this is mainly based on the location of the statue near the Second Pyramid of Giza, which has been connected with Khafra. There is no known inscription from the time of Khafra linking the pharaoh to the Sphinx. The evidence of certain parts of other constructions in the vicinity that appear to be pointing towards the Sphinx have been taken by some as indications that the Sphinx must have been there first, and may have pre-dated the reign of Khafra by two or three hundred years.

An even earlier date was suggested after signs of water erosion were detected on the walls of the Sphinx enclosure. It was said that this water erosion could only have been caused by long and extensive rain. Since Egypt has not had such heavy rainfall since the third millennium BC, that would argue that the Sphinx must have been built earlier, leading to one estimate of 3100 BC or earlier.

442. Which pharaoh was the Sphinx meant to resemble?

Some of those who linked the building of the Sphinx to the time of the Pharaoh Khafra suggested that its face was modelled on Khafra himself. Examinations and detailed measurements by a forensic anthropologist, however, cast doubt on that suggestion.

The author Robert Temple noticed a similarity between the Sphinx's face and that of the Pharaoh Amenemhet II, who reigned from 1929 to 1895 BC. Temple's idea was that the Sphinx was originally a statue of the Jackal god Anubis, but the face was re-carved at a later date.

443. Who destroyed the nose of the Sphinx?

All the evidence suggests that the metre-wide nose did not fall off by accident. Marks on the face suggest that chisels were hammered into the nose and used to pry it off in a deliberate act of vandalism. A wide variety of culprits have been accused of inflicting the damage, ranging from British troops and Egyptian Mamlukes to Napoleon's cannonballs. In one account, a medieval Muslim named Muhammad Sa'im al-Dahr is said to have hacked the nose off to punish Egyptian peasants for making sacrifices to the statue. All we know for sure, from a picture of the Sphinx drawn in 1737, is that the nose was already missing at that date. Which does at least exonerate Napoleon.

444. Is the Hall of Records in Ancient Egypt real?

The Roman writer Pliny the Elder described a cavity around and beneath the Sphinx, which has been linked to an ancient legend of the Hall of Records. This Hall of Records was said to contain a library of papyrus scrolls containing all the knowledge of the Egyptians, including a history of the lost continent of Atlantis.

There have been specific attempts to search for the Hall of Records, and passages of unknown origin have been located around the Sphinx, though Pliny said that the Egyptians believed that it was just the burial chamber of a king.

SPIDERS

445. Is *Aptostichus angelinajoleae* different from *Aptostichus stephencolberti*, or are they the same species?

What do Orson Welles, Harrison Ford, Angelina Jolie, Nelson Mandela, David Bowie and the comedian Stephen Colbert have in common? Answer: they have all had species of spider named after them. Actually Orson Welles had a whole genus of spiders named after him. There is some doubt, however, as to whether the trapdoor spider known as *Aptostichus angelinajoleae* is a different species from *Aptostichus stephencolberti*. The matter should be resolvable by analysing the DNA of both spiders, but since there is no general agreement on what precisely constitutes a species, the question may still remain open.

446. What do catatonic schizophrenics and moulting spiders have in common?

In the 1950s the Swiss pharmacologist Peter Witt began a remarkable series of experiments in which he showed that the patterns of webs spun by a certain species of spider could be altered by giving the spiders drugs. Not only that, but you could tell which drugs the spider had taken by making certain measurements on the resulting web.

One of Witt's early experiments involved the use of hallucinogens, which gave other researchers the idea of giving spiders extracts taken from the urine or the blood of hallucinating psychiatric patients to see whether that had a similar effect on the webs. The most striking result was reported by the Californian neuropsychiatrist Nicolas Bercel in 1959. In Bercel's experiment, a spider that had been fed with a dose of blood serum from a catatonic schizophrenic produced a totally straggly web with no shape or regularity.

Such a lack of shape had only ever been seen before in a web spun by a moulting spider. So to see what was going on, an extract from the body fluids of a moulting spider was then fed to another spider, which again spun a straggly web.

The clear conclusion is that there is something in both the blood of catatonic schizophrenics and the body fluids of moulting spiders that wrecks the patterns of spiders' webs. What it is and how it does the damage remain undiscovered.

See also COFFEE 91

447. What is it that makes spider silk so strong?

It is said that a thread of spider silk the thickness of a thumb would be strong enough to stop a jumbo jet travelling at a cruising speed of 900 kph (570 mph). Bulletproof jackets made of spider silk would only have to be a tenth of the weight of those made from Kevlar. In short, spider silk is one of the strongest substances known to science.

On the assumption that its strength must be due to a protein produced in the spider's spinning organ, genes from spiders have been transplanted into the mammary glands of goats to try to produce goat milk from which an ultra-strong substance could be made. Despite CIA funding and years of research, however, little progress has been made on the production of bulletproof jackets made from goat's milk.

448. How do tiny spider brains store the complex instructions needed to build webs?

Spiders vary in size from large and scary to almost microscopic, and the tiny ones must have really tiny brains. Working with *Anapisona simoni* spiders, which weigh only 1 milligram, William Eberhard of the University of Costa Rica investigated the hypothesis that tiny brains have less room to hold complex web-building information than larger brains.

Comparing the web complexity of the *Anapisona simoni* spiders and their ability to repair deformed webs with those of much larger orb-weaver spiders, Eberhard concluded that the balance of the evidence was that tiny brains are just as good at web-weaving as big brains. Not only that, but the tiniest spiders could design two different types of web, while the bigger ones had only one model.

Yet surely both the memory and processing capacity of tiny brains must be less than those of larger brains, so why is this not reflected in their web complexity?

It is good to love the unknown.

Charles Lamb (1775–1834)

449. Why are the eyes of the *Phintella vittata* spider so special?

In the wide spectrum of all possible electromagnetic radiation, there is a small range of wavelengths that we can detect with our eyes. Thanks to pigments in our retina, we can see wavelengths between about 400 (violet) and 700 (red) nanometres (billionths of a metre). At either end of that range, we have the ultraviolet and infrared parts of the spectrum, which humans cannot see.

We know that the vision of some animals extends into the ultraviolet (UV) range. Several fish, crustaceans, birds and mammals have the ability to see the part of the UV spectrum, known as UVA, which is closest to our own visible light. Between wavelengths of 270 and 320 nm is the UVB range, which is what we need to protect ourselves from when lying in direct sunlight, as UVB is absorbed in a way that can cause damage to our DNA and give rise to skin cancer.

Until recently, it was thought that no animals could see UVB, but in 2007 a team of researchers in Singapore discovered patches on the abdomens of the jumping spider *Phintella vittala* that reflected UVB light and seemed to be used in mating displays. To find out what was happening, an experiment was performed in which two groups of males were separated from females while performing their courtship dance. In one of the groups, the screen separating them included a UVB filter, allowing all light through except that in the UVB range. Result: while normally the females would go wild about the courtship dance, when the UVB filter was placed between them, they showed no interest at all.

Other creatures' eyes are known to be damaged by UVB, so if the female spiders are using their eyes to detect it, which seems very likely, they must in some way be protected. But why, alone of all animals so far investigated, can these spiders look where no other eyes have looked before? One theory is that this UVB detection has evolved as a private form of communication that other species cannot listen in on (so to speak). So what do these spiders have to say to each other that is so private?

450. Why are St Andrew's cross spiders so fussy about their choice of mate?

The St Andrew's cross spider, so-called from its habit of resting in its web with legs outstretched in an X-shape like a St Andrew's cross, has an unlucky sex life. Usually after mating, the male is eaten by the female, and they rarely if ever survive two matings. Despite this, the poor fellows have been seen to have strong preferences over whom they mate with. Experiments in Australia reported in 2004 showed that in the wild, males were strongly attracted to webs spun by laboratory-raised virgin females. Webs from non-virgin females did not seem to attract them at all. Virgin males also showed a preference for virgin females, but males that had previously survived mating did not care who they chose next time. Such a change in mate preference is very unusual, and what it is about the university-educated virgins they like so much also needs explaining.

451. To what extent is the web-planning of spiders a matter of chance, or can they control a web's location precisely?

When you see a spider's web spun in mid-air over some sort of chasm, you may have wondered how they get such a thing started. The answer is that they dangle a thread of silk until it is caught by a gust of wind that carries it over to the other side. With this rope ladder in place, they can then crawl along it, drop more threads, and build up the radii of the web. Then they lay down the sticky spiral of thread which is what catches their food, and right at the end they eat up those original threads that were only put down for construction purposes.

Going back to that first gust of wind, we have no way of knowing what part that plays in determining the precise location of the web.

Does the spider have a subtle feel for wind direction and choose where to dangle its first thread so it will be carried over to the precise point where it wants the web to be, or does it just dangle the thread, happy to build the web wherever the wind blows it?

See also SEX 415

SPONTANEOUS COMBUSTION

452. Has spontaneous human combustion ever occurred?

In his novel *Bleak House*, Charles Dickens gives a vivid description of the death of the rag-and-bone dealer Mr Krook by spontaneous combustion. When critics suggested that people do not suddenly burst into flames, Dickens fervently defended himself, and even added an account of a coroner's inquest supposedly proving that such things can and do happen.

Both historical accounts and even articles in medical journals, especially in the eighteenth century, have supported the idea of spontaneous human combustion, and in recent times bodies have been found after fires that display signs similar to those described in such accounts. One theory, called the 'wick effect', is that an initial flame can split the skin, releasing subcutaneous fat into the clothing, which acts as a wick in a burning candle. Yet no satisfactory explanation has been given of what could cause the initial flame to set off the process. The build-up of internal gases or a high alcohol level have both been suggested, but for spontaneous combustion to occur, either of these would have to be at impossibly high levels. A dropped cigarette still looks the most likely cause.

SQUIRRELS

453. Why do squirrels masturbate?

Nobody thought to ask that question until 2010, when a piece of research on a species of squirrel in Namibia appeared to confound the usual explanations offered to account for such behaviour. There are two standard reasons given for masturbation: it may act as a sexual outlet when a mate cannot be found; or it may (in the case of males) serve to increase the quality of sperm and the chance of fertilization by ejecting old and possibly dysfunctional sperm.

Either of those explanations would predict that masturbation would be most frequent among those who mated less frequently. In the case of the African ground squirrels whose behaviour was monitored in Namibia, however, precisely the opposite was the case: the most frequent masturbators were the males who had the most sex.

The biologist who carried out this research, Jane Waterman of the University of Central Florida, suggested that masturbation may serve as a form of genital grooming, possibly leading to a reduction in the chance of contracting a sexually transmitted infection – but a good deal more research would be needed to confirm that hypothesis.

THE SUN

454. What drives the twenty-two-year solar sunspot cycle?

In 1610, using his newly invented telescope, Galileo made the first observations of the black dots apparently on the Sun's surface known as sunspots. Ever since then, the number of such spots has been monitored and their cause speculated upon. The 'sunspot number' is calculated according to the number of individual spots and the number of groups of spots observed, and this number appears to follow a regular cycle, reaching a maximum every eleven years.

It is now known that sunspots are caused by bursts of intense magnetic activity, and that successive maxima have opposite polarity, so a complete cycle of sunspot activity lasts twenty-two years, but what causes that cycle is unknown. Neither do we understand the effect (if any) of sunspots on the Earth's weather, but that is another matter (→ WEATHER 473).

455. What is the cause of the many unidentified lines in the solar spectrum?

We can tell what elements are present in distant stars by examining the spectrum of light received from them. Any such spectrum includes dark features known as absorption lines, first discovered by the English chemist William Hyde Wollaston in 1802, rediscovered by the German Joseph von Fraunhofer in 1814, and almost fifty years later found to correspond, in many cases, to the emission lines of heated elements. The light spectrum from a star thus offers a celestial fingerprint of the elements present.

Yet Fraunhofer discovered 570 such emission lines in the Sun's spectrum, a figure that has since doubled, leaving a number far greater than the number of elements. The causes of many of the previously unidentified lines are now known, but there remain a large number unaccounted for.

TARDIGRADES

456. Why are the tiny creatures known as tardigrades found almost everywhere on Earth?

Wherever you go, you will not be far from a tardigrade. These microscopic, eight-legged creatures – also known as water bears or moss piglets – can live almost anywhere. No longer than 1.5 mm (3/50th in), they can survive temperatures close to absolute zero or as high as 151°C, and doses of radiation a thousand times the amount needed to kill most other animals. And they can go without water for ten years.

Tardigrades have been found deep in the oceans and high in the Himalayas, in the polar regions and at the Equator. They have even survived trips into space.

There are over a thousand species of tardigrade, but little is known of their evolutionary history. So different from almost everything else, their ubiquitous presence is evidence of a remarkable adaptability, which suggests an important role in evolution. Yet how they have managed to colonize almost everywhere on Earth is a mystery.

UNICORNS

457. Why were the people of the Indus Valley civilization so fascinated by unicorns?

Between 2600 BC and 1900 BC a great civilization flourished along the River Indus in what is now Pakistan and northwest India (→ WRITING SYSTEMS 493). This civilization was only rediscovered in the 1920s by Sir John Marshall, who was astonished by the evidence he found of their highly developed culture. They seem to have been a peaceful people, with no powerful or tyrannical rulers, no wars, no great displays of wealth or elaborate burial rituals, an advanced and efficient method of taxation, a decent sewage system, and a passion for baths – and unicorns. Over a hundred depictions of these fabled beasts have been found on Indus statues and ceramics.

Interestingly, the depiction of the unicorn as a mythological beast is a relatively modern phenomenon. Until the nineteenth century, the general view was at least open-minded as to whether they had existed. In ancient Greek and Roman writings, they appear in natural histories, not in mythology, with both Aristotle and Pliny the Elder indicating that they were to be found in India.

So were the Indus unicorns real? Sadly, no unicorn skeletons or fossils have been dug up in the region, but if they were not real, where did they get the idea from, and how did it establish itself so strongly in their culture?

THE UNIVERSE

458. Is there life elsewhere in the universe?

There are thought to be about 200 billion galaxies in the observable universe, each of which has about 200 billion stars. How many planets orbit how many of these stars is not known, but with numbers as vast as these, there seems to be a good chance that at least one of the 40,000,000,000,000,000,000,000 stars has a planet on which life has evolved. On the other hand, the more we look at life on Earth, the more we realize how delicately balanced the physical and chemical features have to be to allow life to come into existence.

Stars themselves, however, do not go on forever, and the right conditions for life may only exist for a short period of their existence, so while there may once have been life elsewhere in the universe, we may have missed it. With that number of chances, however, it seems very likely that there is some sort of life somewhere out there. Which brings us to the next question...

459. Is there intelligent life elsewhere in the universe?

In 1961, when the serious search for extraterrestrial intelligence was in its infancy, the American astrophysicist Frank Drake produced a famous equation to predict the chance of our making contact with intelligent life elsewhere in our galaxy, the Milky Way. His formula took into account the rate at which new stars are created, and the chances that they will have planets, that those planets will have the conditions to support life, that life will actually have formed, that it will develop intelligence, that its technology will be good enough to send out or receive radio signals, and that its civilization will not have died out. Multiply together all these factors, and you have the chance of our making contact.

The trouble with Drake's equation is that almost all the terms in it are a matter of guesswork, and the estimates it comes up with for the number of Milky Way civilizations we may contact range from about ten to around 10 billion. There is also, of course, the possibility that many of these may be far more advanced than we are, and may not want to talk to us anyway.

Given that the Milky Way is only one of some 200 billion galaxies, however, the chance of intelligent life somewhere out there, even if we have little hope of contacting it, must be very high – according to Drake's equation anyway. But we shall not know until we get a reply to one of our transmissions, or pick up one of theirs.

460. What is the Omega value of the universe?

In his book *The Phenomenon of Man*, written in the 1930s but not published until 1955, the French Jesuit priest, palaeontologist and philosopher Pierre Teilhard de Chardin (1881–1955) came up with the concept of an 'Omega Point' for the universe. If Alpha, the first letter of the Greek alphabet, represents the Creation then Omega is the end, which he defined as the state of maximum complexity towards which the universe is evolving.

When the Big Bang theory began to establish itself, Teilhard's Omega Point led to the idea of an Omega value for the universe that would predict its ultimate fate. The Big Bang led to all the matter in the universe sharing a massive amount of kinetic energy, which has been fuelling its expansion ever since. On the other hand, the gravitational pull of all the mass in the universe is working against this expansion by pulling everything back together again. The idea of the Omega value is to provide some sort of ratio between this kinetic energy and this mass, a ratio that will determine the ultimate fate of the universe. If Omega is less than 1, then the universe will continue expanding forever; if it is greater than 1, the expansion will come to an end before going into reverse and ending in a Big Crunch; if it is exactly equal to 1, then it will eventually approach a steady state.

Until recently, ideas of the matter-density of the universe predicted an Omega value very close to 1, leaving its fate still a three-horse race. Then came the theories of dark matter and dark energy, which have affected estimates of both the density and the expansion rate, making it more difficult than ever to predict what will happen at Teilhard's Omega Point. The smart money these days seems to be on an Omega less than 1 and eternal expansion, but that could all change with more cosmological discoveries.

461. What is the shape of the universe?

The question of the Omega value (→ 460) and the ultimate fate of the universe is tied in with its geometry and the question of whether Pythagoras' theorem works over vast distances. (Just in case you don't recall it, Pythagoras' theorem states that in any right-angled triangle, the square of the hypotenuse is equal to the sum of the squares of the other two sides.) Think of measurements made on the Earth's surface. Over short distances, Pythagoras works pretty well because the Earth is almost flat and so obeys the laws of Euclidean geometry. Over large distances, however, the curvature of the Earth comes into play and Pythagoras fails us.

Now try to imagine the same thing over the vast expanse of space. Considerations of the universe's expansion have led to the conclusion that if Omega is equal to 1, then the Universe is flat (in the sense that Pythagoras' theorem gives the right answers); if Omega is greater than 1, then space has positive curvature, rather like the surface of a sphere (though here we are actually talking about a hypersphere); and if Omega is less than 1, it has negative curvature and it is like taking measurements on a saddle.

Whatever the basic geometry, there is still an argument over the overall shape of the universe, with the debate focusing on whether it has been expanding at the same rate in all directions. Just as the Earth is not a perfect sphere but is a little squashed at the poles, the universe might not be spherical, but an ellipsoid. That would help solve some of the problems associated with the microwave background radiation (→ 462), but would introduce another problem of explaining how it got that way in the first place.

462. Why is the temperature of the microwave background radiation apparently uniform everywhere in the universe?

The cosmic microwave background radiation is a sort of afterglow of the Big Bang, the result of the initial massive surge of energy and heat being chilled by the expansion of the universe. Readings from NASA's Cosmic Background Explorer (COBE) satellite indicate a remarkable uniformity in this background radiation in every direction. Its temperature scarcely varies from a figure of about 2.76 degrees above absolute zero wherever one looks. Indeed, measured variations have only been around 0.001 per cent of the average figure.

This may seem all neat and tidy, but it is at variance with theories of how matter developed. Current theories predict changes in temperature according to the amount of matter in different regions. On the one hand, the almost perfect uniformity of cosmic background radiation is one of the strongest pieces of evidence supporting the Big Bang theory, but on the other hand, finding a way to account for its tiny fluctuations in terms of variations in the distribution of matter in the universe still poses great difficulty.

463. Is the universe finite or infinite in size?

The idea of a Big Bang and the limitations of the speed of light impose a limit on the size of the visible universe, but if the universe is expanding, that size is always increasing. Is there any limitation on the maximum size it could occupy?

Suppose a spacecraft takes off from Earth and travels in a straight line through space. Will it go on forever into the infinite universe, or will it, as some versions of the universe's geometry claim, eventually come back to its starting point? Such a finite universe is not out of the question, and claims have been made that it might explain ripples in the cosmic microwave background.

464. If string theory is right, how would we detect the influence of the other seven dimensions of the universe on our three-dimensional slice of it?

In Edwin Abbott Abbott's wonderful mathematical fantasy *Flatland* (1884), we meet a world of two-dimensional beings living on a flat plane who are visited by a sphere passing through their world. At first the sphere appears as a dot, then grows to a small circle, gradually increasing in size, but getting smaller as it continues its way through, before fading to a dot again then disappearing. Whether an analogous effect could be produced by a ten-dimensional inhabitant or piece of material of the ten-dimensional string-theory universe (→ PHYSICS 359) in passing through our three dimensions is an intriguing question.

Or perhaps we ourselves are in fact ten-dimensional creatures with the misfortune to be saddled with brains and sensory organs that can only recognize three of them. Spotting signs of the extra dimensions, however, seems to be a more difficult task for us than it was for the squares and triangles of Flatland.

465. Is the universe eternal?

Even Stephen Hawking seems to change his mind on this one. Not so long ago, he and other cosmologists were adamant that the Big Bang was the start of everything. Even asking 'What happened before the Big Bang?' was not a meaningful question. Now they are not so sure. And in the same way, 'What happens after the end of time?' may also be an acceptable question.

Whether or not the universe will go into reverse and end in a Big Crunch, we know that matter decays into smaller and smaller particles. With black holes sweeping up everything in their wake, one picture of the distant future has everything disappearing into a massive black hole containing all the energy in the universe.

In the past it was thought that nothing whatsoever could ever emerge from a black hole. Now it is believed that they can emit certain types of radiation, and radiation is energy, and energy is matter. This allows for a picture of an everlasting universe, acting on an unimaginably long time scale, pulsing between Big Bang creations and Big Crunch endings, into massive black holes which eventually explode again into another Big Bang. But we will have to wait a very long time indeed to know whether that scenario is correct.

466. How will it end if it isn't?

The old Norse legends had Ragnarok, other religions have Judgement Day or the Apocalypse, Wagner had *Götterdämmerung*, the twilight of the gods. Among more scientific theories, we have the Big Freeze, in which the temperature of everything approaches absolute zero. Alternatively, there is the theory of the Big Heat, where the laws of entropy bring everything into a state of maximum confusion and uniform temperature, with no free energy left to sustain life or any form of motion. The Universe would just stop. Then there is the Big Rip, when dark energy (possibly a kind of anti-gravity) tears everything apart, and the Big Crunch, where the universe collapses in on itself after the momentum of the Big Bang comes to an end and gravity reverses the process. Take your pick.

See also COSMOLOGY 106–14, PHILOSOPHY 367, QUANTUM PHYSICS 399–407

VENUS DE MILO

467. Who was the sculptor of the Venus de Milo?

Perhaps the most famous sculpture in the world, the Venus de Milo, was discovered in a cave on the Greek island of Milos in 1820. It was originally attributed to the master sculptor Praxiteles, which would have meant that it was made in the fourth century BC, but later a plinth was found that was said to fit the sculpture exactly and which included an inscription saying: 'Alexandros son of Menides, citizen of Antioch on the Maeander made this.' From the style of the lettering, this inscription would have dated the sculpture to around 100 BC, but some have suggested the inscription was not part of the original plinth but was only added later. To add to the confusion, the plinth quickly went missing after the statue arrived in Paris in 1821, and the only evidence we now have of the inscription is in a drawing made of the piece.

468. Who was the Venus de Milo meant to depict?

The general view is that she is most likely to be the goddess Aphrodite, whom the Romans called Venus and who was often portrayed half naked. There is even a suggestion that a missing piece of marble may have been an apple she was holding, making the statue suggestive of the Judgement of Paris. Others maintain that she is not Aphrodite but the sea goddess Amphitrite, who was venerated on Milos, where the statue was found.

469. What happened to the arms of the Venus de Milo?

In his memoir *Two Voyages to the South Seas*, the French explorer and sea captain Jules-Sébastien-César Dumont d'Urville (1790–1842) gave a thrilling account of his struggle with Greek brigands when transporting the Venus de Milo from Greece to France. In the fracas, he says, the statue was roughly dragged across rocks to the ships, resulting in both arms breaking off. The French sailors, eager to get away from the brigands, refused to return to search for them.

This whole story, however, appears to be pure fabrication, as other accounts and sketches indicate that the arms were absent when the statue was found. In either case, the missing arms have never been located. Nothing at all is known of what happened to the statue before or after it was hidden in the cave where it was found.

WATER

470. Where does the Earth's water come from?

This has always been a bit of a puzzle, as the Earth must have been much too hot in its early days to sustain any water on its surface. The usual theory is that water was deposited on the Earth from collisions with comets and asteroids in the planet's youth. Some investigations into the water content of such bodies has lent support to that idea by showing them to have much the same ratio as water on Earth of 'normal' hydrogen to deuterium (an isotope of hydrogen with a neutron in its nucleus). Other recent studies, however, notably of Comet Hale-Bopp, have led some to question the whole theory of the possible cometary origins of the Earth's water.

In 2007 Japanese scientists came up with an alternative hypothesis. In the early days, they suggest, the Earth had a great deal of hydrogen in its atmosphere, and this reacted with oxides in the Earth's mantle to form water. The heavy hydrogen cloud, they maintain, could also explain why the Earth's orbit round the Sun changed from an original elliptical shape as predicted by theory into the almost circular path we see today. While initially the water produced from their model might have been expected to be low in deuterium, their calculations also explain how it would have changed over time to the present amount.

471. Where did the water on the Moon come from?

In 2009 NASA crashed a lunar satellite into a crater on the Moon and announced that to their surprise they had found 'irrefutable' signs of significant quantities of water. But where did it come from? The scientists involved made all sorts of suggestions including solar winds, asteroids, comets, gases from the Moon's interior and grains of ice carried by intergalactic clouds. Or it may have come from the Earth. Or, in short, it might have come from anywhere.

Further lunar experiments are planned to get to the bottom of the mystery – or, as one scientist put it, to 'stuff the water back into the crater'.

472. How do ice crystals form?

We think of ice as a single entity: it's just frozen water after all, and water is water, right? Well, not really. Fifteen different forms of crystalline water have been identified. These different forms appear to result when freezing takes place in different conditions of temperature and pressure, but what precise conditions lead to which crystalline forms, and how changes in these conditions lead to differences in the geometry of the hydrogen bonds in the water molecules, are not yet completely understood. Only when we have achieved that understanding will we be able to explain the geometry of snowflakes.

473. What starts the process of ice formation in clouds?

The formation of ice droplets in a cloud has a highly significant effect on climate, influencing both the cloud's cooling effect in reflecting away the Sun's rays and its warming effect in trapping the heat radiating from the Earth. Yet we have very little understanding of how these ice droplets form.

When water is above a temperature of 0°C, it is liquid; when it is below –38°C, it will always be frozen; but between 0°C and –38°C it can be either. In this range, a water droplet needs a nucleus around which to freeze. Soot, mineral dust, meteoric particles, bacteria or pieces of ice may serve as this nucleus, but to what extent each different type of material performs this function, whether they are effective at different temperatures, and whether they lead to different crystalline formations of ice are uncertain. In 2009 a number of papers appeared supporting the view that biological particles such as bacteria play a strong role in the process, but research in Germany in 2010 claimed that dust was much more important. Until this is sorted out, we shall not know what to blame for the rain – which is, of course, what we notice when the ice crystals grow heavy enough to fall from the clouds, melting on their way down to the ground.

474. What is so special about water?

When two positively charged ions of hydrogen attach themselves to one negatively charged ion of oxygen to form a molecule of water, an additional factor that keeps them together is something called a hydrogen bond, which is the attractive force of the opposite charges.

Much of what we identify as the unusual properties of water seems to be dependent on the nature of the hydrogen bond: the fact that so many substances dissolve in water; the fact that water expands as it freezes; the surface tension of water. Life as we know it would be impossible without these special properties. Just think of the watery origins of life. If ice were heavier than water, it would sink to the bottom, leaving water at the top to freeze and sink again, until all life was extinguished from any body of water in sub-zero conditions. As it is, a layer of ice stays at the top, insulating the rest of the lake and allowing life to go on. And most of the salts and other chemicals that have led to the development of life have relied on water's solvent properties to get where they need to go.

Water is one of the most commonly encountered yet most unusual and least understood chemical compounds on Earth, and most of this seems to be due to the strange properties of the hydrogen bond.

WEATHER

475. How do tornadoes and tropical cyclones form?

'Cyclone' is a wonderful word. It signifies 'turning in a circle', just as the one-eyed Cyclops who gave Odysseus so much trouble might have done. Tropical cyclones are violent storms, including hurricanes and typhoons, which at their most dramatic exhibit the truly cyclonic, twisting behaviour which characterizes the tornadoes of North America. Such powerfully twisting storms seem to need several factors for their formation: warm water temperatures sustained to a considerable depth, high humidity, air rapidly cooling with height; consistent wind speed and direction, but otherwise disturbed atmospheric conditions. They also seem to need to be at least five degrees of latitude away from the Equator in order for the Coriolis force to deflect winds and create the turning effect. There is clearly something about them that we do not understand.

476. Is accurate long-range weather forecasting possible?

Short-term weather forecasting is based on taking measurements of current weather conditions, assessing where clouds and air masses are heading, and trying to work out what they will do when they get there. As computer systems become larger, and more detailed observations are fed into them, the results may become more accurate, but it seems that this accuracy is limited to a period of about five days.

The trouble is, weather is chaotic. When the American mathematician and meteorologist Edward Lorenz introduced the subject of chaos theory in 1972, he entitled his seminal paper 'Does the Flap of a Butterfly's Wings in Brazil Set Off a Tornado in Texas?' The whole idea of chaos theory is that it is possible to have a system whose behaviour is in principle totally determined by Newton's laws of motion, yet even the tiniest changes in the initial conditions can lead to completely different consequences. One set of conditions may lead to fine weather in Texas, but if you add one flap of a butterfly's wing in Brazil, you end up with a tornado. And even if your equipment is sensitive enough and your computer powerful enough to include the wing flaps of every butterfly on Earth, then the exhalation of an ant could still throw everything way off course.

One approach to long-range forecasting is to incorporate the unpredictability of chaos into computer simulations and run the data through several times, with slight changes to the initial figures.

If they all give the same result, one can be confident it is right (as long as the computer model is basically sound); but if they give different results, all one can do is offer a probabilistic estimate of the chance of different types of weather. And that, until we understand the weather much better and can better predict sea currents, air movements and their effect on one another, may be the best we can do for long-term forecasts.

477. What effect, if any, do sunspots have on our weather?

For more than a century, an argument has continued over the effect on our weather of sunspots (→ THE SUN 454). At various times, correlations have been published between the amount of sunspot activity and temperatures on Earth. The most striking example comes in data from the period 1645–1715, a period known as the Maunder Minimum after the English astronomer Edward Maunder (1851–1928) who first drew attention to it. This was a period of very low sunspot activity, which coincided with the severest part of the spell of prolonged cold weather known as the Little Ice Age. A link has long been suspected between the two events, but no convincing mechanism has been suggested to explain how sunspot activity could affect our weather. Other attempts to link sunspot activity to weather conditions have also failed to convince.

478. What is the cause of ball lightning?

Over the centuries there have been many accounts of glowing balls of fire appearing in the sky, usually during thunderstorms. These balls move through the air at a few metres per second, then either disappear or explode leaving a smell of sulphur. For long, such phenomena were dismissed as fantasies or hoaxes, but more recently the idea of ball lightning has been taken more seriously.

Vaporized silicon, aerodynamic vortices, large electrostatic charges and even black holes have been advanced as possible explanations, and scientists have produced in the laboratory effects that match accounts given of ball lightning. But there is still no way of knowing whether any of these laboratory-created phenomena are actually the same as those that take place in nature.

479. Why are noctilucent clouds becoming more frequent?

Noctilucent clouds, composed of tiny crystals of ice, are the highest clouds in the Earth's atmosphere. They are only visible when they reflect sunlight, especially as night draws on, which gives them their name, meaning 'night-shining'. Noctilucent clouds were first observed in 1885, two years after the massive eruption of Krakatoa, causing a suspicion that their existence may in some way have been due to volcanic debris in the atmosphere. Or it could be that the spectacular sunsets caused by the debris made people take more notice of what was going on in the sky.

In recent years noctilucent clouds have increased in frequency, which has led to a suggestion that their appearance is somehow connected to climate change. However, no mechanism has been suggested for such a link.

480. Will it rain next week?

The main thing we want to know from the weather forecast is whether it's going to rain. Up to about five days ahead, weather forecasts have a fair chance of being right (→ 476). A month or more ahead, a reasonable prediction of the chance of any particular type of weather may be made on the basis of past history. One week ahead falls uncomfortably between the two types of forecast. Frankly, your guess is as good as that of the next meteorologist.

See also CLIMATE 87–8

WORMS

481. What is the function of the hairs and bristles on the sperm of worms?

Making videos of tiny worms having sex under a microscope may seem an unusual way to spend one's time. Yet biologists in Switzerland have been doing just that with flatworms of the *Macrostomum* genus, their aim being to solve an old problem regarding the evolution of bristles and other adornments on the sperm of worms.

Macrostomum is an interesting little worm, about the size of a comma. They are completely hermaphroditic, having both male and female sex organs. When they mate, they curl together like two interlocking letter Cs, and the male organ of each enters the female organ of the other.

What the researchers were surprised to observe, however, was that after mating, each tried to remove the other's sperm by sucking it out. They conjecture that this is a method of sexual selection, choosing who they want to father their children not before mating, as is the common practice with many creatures, but after. They also suggest that the role of the bristles on the sperm of *Macrostomum* is to make it more difficult to suck out. Only making more videos of mating worms, of this and other genera, will confirm whether this is truly what happens.

> In other words, apart from the known and unknown, what else is there?
>
> Harold Pinter (1930–2008)

WRITERS

482. Who wrote *Beowulf*?

The epic poem *Beowulf* is the most important surviving work of Anglo-Saxon literature. Known only from a single manuscript called the Nowell Codex, it is a tale of heroic dragon-slaying exploits so potent that a translation of its 3,182 lines into modern English won Seamus Heaney the Whitbread Book of the Year prize in 1999. Yet we know almost nothing of the origins of *Beowulf*. The best that can be said is that it was written somewhere between the eighth and eleventh centuries. As for the identity of the writer, we have no idea at all. The story may, however, have existed for some time, being passed down in the oral tradition, before finally being written down.

483. Who wrote *Sir Gawain and the Green Knight*?

Unlike *Beowulf*, it is at least possible to draw some deductions about who wrote *Sir Gawain and the Green Knight*, and even to hazard a guess. A take recounting an adventure of King Arthur's youngest knight, who accepts a challenge from an unknown knight dressed in green, *Gawain* was written in the fourteenth century in a dialect of the northwest Midlands, a fact that is assumed to give a clue to the author's origins. Literary historians have also deduced that the author had a knowledge of French, Latin and theology, but was probably not a theologian himself.

The only name anyone has come up with is that of John Massey of Cotton in Cheshire, who is known to have written the poem *St Erkenwald*, which has some stylistic similarities with *Sir Gawain*. Massey's dates, however, are unsure, and there is even some doubt about whether he was alive when *Sir Gawain* was written.

484. Was Geoffrey Chaucer a rapist?

This question relates to an incident in the life of the author of *The Canterbury Tales* that has been a matter for intense speculation. On 4 May 1380 a woman called Cecily Chaumpaigne presented to the Chancery of Richard II a 'deed of release', which was copied into the court records. In it, she 'released' the poet Geoffrey Chaucer from 'all manner of actions such as they relate to my rape or any other thing or cause'. And that is essentially all we know of the case, except for a further suggestion that Chaucer paid Cecily £10.

Quite apart from the lack of any other information, the document was written in Latin, which has led to much argument about whether a rape was involved or not. The phrase *'de raptu meo'* could, according to some, have referred not to a rape, but more likely to a kidnapping or abduction.

To add to the mystery, the issuing of a 'deed of release' means that it was a civil case, not a criminal one. This, together with the £10 payment, suggests that Cecily did bring, or threaten to bring, a claim against Chaucer, and that he settled out of court. It is likely that we will never find out the full story behind these intriguing hints.

485. Who was the 'person from Porlock' who disturbed Coleridge in his writing of 'Kubla Khan'?

Samuel Taylor Coleridge claimed that the entire text of his poem 'Kubla Khan' (completed in 1797) had come to him in a dream, which he hastened to put down on paper on awakening. However, as he explained in an introduction when he first published the poem in 1816, a terrible thing then happened. (He refers to himself in the third person.)

> *On awakening he appeared to himself to have a distinct recollection of the whole, and taking his pen, ink, and paper, instantly and eagerly wrote down the lines that are here preserved. At this moment he was unfortunately called out by a person on business from Porlock, and detained by him above an hour, and on his return to his room, found, to his no small surprise and mortification, that though he still retained some vague and dim recollection of the general purport of the vision, yet, with the exception of some eight or ten scattered lines and images, all the rest had passed away like the images on the surface of a stream into which a stone has been cast, but, alas! without the after restoration of the latter!*

So who was this 'person from Porlock'? Or was he just an invented excuse? One suggestion is that it was Coleridge's doctor, P. Aaron Potter, who regularly supplied the poet with laudanum (tincture of opium).

486. What did Edgar Allan Poe die of?

On 3 October 1849 the forty-year-old Edgar Allan Poe was found in great distress and a state of delirium on the streets of Baltimore, Maryland. He was taken to Washington College Hospital, where he died four days later, having never in that time been coherent enough to explain what had happened.

Poe had last been seen on 27 September, when he left Richmond, Virginia, on his way home to New York. No clear evidence of his whereabouts is known until he was found in Baltimore. During his four days in the hospital, he was kept in an area reserved for drunks, and was denied any visitors. On the night before he died, he is said to have repeatedly shouted out the name 'Reynolds'.

All medical records, including the death certificate, have been lost. Newspaper obituaries gave the cause of death as 'congestion of the brain' or 'cerebral inflammation', both being common euphemisms for alcoholism. Yet the doctor who treated him said there was not a trace of the smell of alcohol on his clothes or breath. Other suggestions include heart disease, epilepsy, syphilis, meningeal inflammation, cholera and rabies.

487. Why was Poe wearing someone else's clothes when he died?

To add to the mystery, when he was found in Baltimore, Edgar Allan Poe was apparently dressed in someone else's clothes, to judge from their shabbiness, which would have been most unlike him. The owner of the clothes was never identified, but has led to speculation that he might have been a victim of 'cooping', an electoral vote-rigging scam that involved seizing and drugging someone off the streets and turning up with them to vote at a number of polling stations. However, Poe was well known in Baltimore, and even shabby clothes might not have stopped him being recognized.

488. Why did Darwin delay so long before publishing *On the Origin of Species*?

Charles Darwin returned from his long trip on the *Beagle* in 1836, and by the end of 1838 told friends that he had the plans in place for his work on natural selection. Yet twenty years later, in 1858, he told Alfred Russel Wallace that he was not yet ready to publish. Darwin's great work, *On the Origin of Species by Means of Natural Selection*, did, however, come out the following year, in 1859, and was an immediate success. So why the delay of two decades?

One natural explanation could be Darwin's fear of upsetting his many friends in the clergy, or even his devout wife, Emma. He may even have feared religious persecution. All these may have played a part, but the most probable explanation could be that he was aware of the revolutionary impact his ideas would have, and was therefore determined to amass his evidence and polish his arguments to make sure he got it right.

The science historian and Darwin specialist John van Wyhe has suggested that the question needs no answer, as in Darwin's time twenty years would not have been seen as particularly long for a work of such magnitude. All the same, Darwin must have known of others' interest in the topic, and non-publication always ran the risk, even in the Victorian era, that someone else would get in first. There is good reason to believe that what eventually prompted Darwin's final decision to publish was a letter from Wallace outlining his own version of the theory of natural selection, with the implication that he was ready and prepared to publish it himself.

489. Was Émile Zola murdered?

Accident, suicide or murder? When in 1902 the sixty-two-year-old French novelist and political activist died of carbon monoxide poisoning caused by a blocked chimney in his home in Rue be Bruxelles, in Paris's seventh arrondissement, any of the three seemed possible. When Zola, in 1898, had written his famous front-page open letter to the president of the French Republic, headed with the words 'J'accuse!' and charging the president with anti-Semitism and obstruction of justice in the Dreyfus case, he had made some powerful enemies. There had even been attempts on his life. His enemies, not unnaturally, celebrated his death and said, rather unconvincingly, that Zola had committed suicide on discovering that Dreyfus had been guilty after all (which turned out not to be the case).

Many years later a Parisian roofer, on his deathbed, is said to have confessed to blocking Zola's chimney 'for political reasons'. Yet that does seem a rather elaborate way of murdering someone, and the roofer did not say on whose orders he was acting.

490. Whatever happened to Ambrose Bierce?

Ambrose Bierce – journalist, satirist and author of *The Devil's Dictionary* – was seventy-one years old when he left Washington, DC, in October 1913 for a tour of Civil War battlefields. After that he is said to have joined Pancho Villa's army as an observer of the Mexican Revolution, and was in Chihuahua on Boxing Day 1913 when he wrote a letter to his friend Blanche Partington. He signed off with the words, 'As for me, I leave here tomorrow for an unknown destination.' He was never seen again.

One tale is that he was shot by a firing squad in Sierra Mojada, but others say he never went to Mexico at all. Despite several attempts, no trace of his movements has ever been found.

491. What became of the first draft of T. E. Lawrence's *The Seven Pillars of Wisdom*?

In 1919 Lawrence of Arabia had almost completed his first draft of *The Seven Pillars of Wisdom*, most of which he had written while attending the Paris Peace Conference. Around Christmas, back in England, he took the manuscript with him on a train journey that necessitated a change at Reading. While waiting for his connection, however, his briefcase, with the precious manuscript inside, went missing.

Some accounts say he left it on the train, others that he left it in the station buffet; some say it was stolen. All that remained in Lawrence's possession was a typescript of a few early chapters. The rest he rewrote from memory, and the book was published in 1921. The missing manuscript, and the briefcase, were never found.

> Ignorance gives one a large range
> of probabilities.
>
> George Eliot (1819–1880)

492. Why did Agatha Christie disappear for eleven days in December 1926?

Towards the end of 1926, the crime writer Agatha Christie, then aged thirty-six, was told by her husband Archie that he was in love with another woman and wanted a divorce. Not unnaturally, they quarrelled and on 8 December he left their house in Sunningdale, Surrey, to spent the weekend with his lover. That same evening, Mrs Christie left the house, leaving a note for her secretary saying that she was going to Yorkshire.

News of the disappearance mobilized Agatha Christie's army of fans, and a nationwide hunt was set up to find her. It took eleven days before she was identified at a hotel in Harrogate where she had checked in as 'Mrs Teresa Neele', using the same surname as her husband's lover. She never gave any account of what she had been doing for those eleven days, and suggestions have ranged from a nervous breakdown or a publicity stunt to an attempt to embarrass her husband or even have him framed for her murder.

See also SHAKESPEARE 423–7

WRITING SYSTEMS

493. Was the strange script of the Indus Valley civilization an unknown language, an accounting system, or what?

Serious excavations of the Indus Valley civilization, which thrived in what is now Pakistan and northwest India between 2600 and

1900 BC (→ UNICORNS 457), began only in 1920. But even before that, in the 1870s, a picture of a seal from the Indus people had been published, clearly showing their curious hieroglyph-like inscriptions. Since then, over 4,000 such seals or tablets have been found, providing a wealth of material but no long texts that might offer some information about the language. In fact, the typical tablet contains only five symbols, raising the possibility that it is not a language at all, but some sort of counting or messaging system. The longest inscription contains only seventeen signs.

Over the years, many have claimed to have deciphered the script, but no two experts seem able to agree even on what group the language of the Indus people might have belonged to, or what their symbols might mean. The only tentative conclusion that does have majority support is that the Indus script was written from right to left.

494. What did the symbols of the alphabet known as Linear A mean in ancient Crete, and what was the Minoan language they may have transcribed?

When Sir Arthur Evans excavated the palace of Knossos in Crete in the early 1900s, he discovered numerous clay tablets containing inscriptions in three different scripts. One consisted of hieroglyphs, the others he named Linear A and Linear B. In 1952 the English classical scholar Michael Ventris showed that Linear B was used to write an early form of Greek, but Linear A, though it contains some of the same symbols as Linear B, has never been decoded.

When our knowledge of the Linear B symbols is applied to Linear A, it produces words that appear to be unrelated to any known language. One suggestion is that this is the lost Minoan language, spoken in Crete before invasions by the Greeks around 1450 BC. Attempts to reconstruct this language and decipher the Linear A tablets have met with limited success at best. This may cast doubt on the assumption that the symbols shared by Linear A and Linear

B have the same syllabic values in each language, which would mean we would have to go back to square one in our attempts at decipherment.

As for the language that may have been behind Linear A, there is some evidence that it was Indo-European, or a pre-Indo-European precursor, though not enough to be sure. A form of pre-Greek has also been suggested, as has Luwian, an extinct language spoken in Anatolia. An archaic form of Phoenician is another idea, while Indo-Iranian and Tyrrhenian have also been proposed.

See also ANCIENT HISTORY 10

495. What is the language of the Voynich Manuscript, or is it an elaborate fake?

The Voynich Manuscript has been described as 'the world's most mysterious manuscript'. Others consider it likely to be one of the world's greatest hoaxes.

The beautifully illustrated manuscript is written in an unknown script and has defied all attempt by cryptologists to decode it, although the illustrations suggest it may have been some kind of pharmacopoeia or medical text. The manuscript – named after the book dealer Wilfrid M. Voynich, who acquired it in 1912 – has been carbon-dated to the first half of the fifteenth century.

The text consists of 170,000 characters, separated by narrow gaps, but with longer gaps dividing them into something resembling words. There are about twenty to thirty distinct characters that occur frequently, with about another dozen that make rare appearances. Statistical analysis suggests that the letter frequency resembles that of natural languages, but there are no very long or very short words. Its lack of clear resemblance to any particular language suggests a code, but, if that is the case, nobody has been able to crack it. It is not known where the manuscript originally came from, but it was in the possession of an alchemist in Prague in the seventeenth century.

If the Voynich Manuscript is a forgery, it is brilliant both in design and execution, which must have demanded the use of genuine fifteenth-century writing materials. If it is not a forgery, it is extraordinary that it has resisted every attempt – using all our knowledge of fifteenth-century coding methods and twenty-first-century computing power – to decipher it.

496. Where did our alphabetical order come from?

The Egyptians had hieroglyphs, which began as a form of pictogram and later came to represent sounds rather than objects. By 2700 BC the Egyptians had a sort of alphabet of twenty-two hieroglyphs representing the consonants in their language. The Phoenicians, around 1700 BC, were the first to come up with an alphabet that was beginning to look a little like ours, and the Greeks and Romans developed their alphabets from the Phoenician, keeping much the same order of letters as the Phoenicians had, but introducing letters of their own.

The Roman schoolteacher and grammarian Spurius Carvilius Ruga is credited with inventing the letter G, to avoid the confusion of C being used both for the 'k' and 'g' sounds. But why were the letters of the Roman alphabet (the one we use today) placed in a particular order, beginning with A?

The only answer seems to be that the Romans copied the Greeks, and the Greeks copied the Phoenicians, and nobody knows the process by which the Phoenicians arrived at their order.

See also EASTER ISLAND 154

Ignorance of one's misfortunes is clear gain.

Euripides (*c.* 450 BC–406 BC)

YETI

497. Is there any truth in the story of the abominable snowman?

The word 'yeti' is said to come from two Tibetan words meaning 'rocky place' and 'bear'. This bear from a rocky place captured the Western world's imagination when the Everest Reconnaissance Expedition of 1921 returned with tales of mysterious footprints in the snow. The Sherpa guides had said these must be those of the 'Wild Man of the Snows', for which they used an expression that was mistranslated as 'abominable snowman'. And the legend of the abominable snowman was born.

The yeti, however, had been part of local beliefs since before Buddhist times. He had been depicted as an apelike creature, making a whistling sound and carrying a large stone as a weapon. Around 1950, interest in the yeti peaked, following several claimed sightings and some unexplained footprints. The *Daily Mail* even sponsored a Snowman Expedition to find it. As recently as 2007 and 2008, more footprints have been sighted, and suspected samples of yeti hair found, but nothing has been confirmed, and the elusive beast has remained hidden. Some cryptozoologists like to think that the yeti is some kind of apeman or missing link, while the scientific community is inclined to believe that the footprints and sightings are more likely to have been of a bear – albeit a ' bear from a rocky place'.

498. What happened to the supposed finger and thumb of a yeti that the actor James Stewart helped to smuggle out of India in 1959?

In 1959 the Irish-born mountaineer Peter Byrne visited a lamasery in Pangboche, Nepal, where he hoped to examine something that was supposed to be the hand of a yeti. Having come prepared, Byrne stole a finger and thumb from the hand, replacing them with a human finger and thumb that he had brought with him. The supposed yeti parts were smuggled into India where Byrne bumped into the actor James Stewart and his wife Gloria, who agreed to help smuggle them to England. They are said to have wrapped them in underwear and carried them in their hand luggage. Sadly, this was in the days before DNA identification, so scientific tests on the digits were inconclusive. The samples subsequently vanished and have never reappeared. In May 2011 a replica of the missing yeti hand was returned to the monastery in Nepal as part of a 'Return the Hand' campaign. But the original yeti hand has still not turned up.

ZYMOLOGY

Zymology is the study of fermentation, in which, as you probably know, yeast plays an important part. Of course, the most interesting question of all about yeast is this:

499. Can yeast think?

Writing about the evolution of the brain in the *Sydney Alumni Magazine* in 2009, the molecular neurobiologist Professor Seth Grant posed the question 'Can yeast think?' His answer was rather surprising: 'At some level the answer to this is yes. Yeast cells make decisions in response to changes in their environment,' he said, and went on to explain how they detect changes in their environment and alter their growth or behaviour accordingly. That would not, of course, be considered 'thinking' by most people, but his view is supported by the discovery that the proteins used by yeast to make such changes are the same as some proteins found in the synapses of the human brain. So the origins of our own brain evolution may lie in the ancestral proteins of unicellular fungi such as yeast.

AND FINALLY...

500. Will there always be things we don't understand?

It certainly looks that way. The more we understand the more we realize we do not know – and the more we are inspired to find out more. As Benjamin Franklin put it: 'The doorstep to the temple of wisdom is a knowledge of our own ignorance.'

I hope this book will have given a useful lift towards that doorstep.

501. If there comes a time when we know everything there is to be known, will we recognize it?

...or will there come a time when the only thing we do not know is that we know everything?

ACKNOWLEDGEMENTS

Apologies and huge gratitude are due in about equal measure to all the academics and other experts whom I approached in the course of writing this work with requests that they tell me everything they didn't know. Responses varied from fascinating outpourings of ignorance to blank looks of incomprehension. I have decided against even trying to compile a list of all who helped, even by vaguely pointing me in promising directions, partly because the complete list would be too long, but also because I am sure I would offend someone by leaving them out. A particularly delightful Japanese lunch with Norman Lebrecht does, however, linger in the memory, and revealed to me the full mystery of Scriabin's shaving cut. For the rest of my anonymous guides, you may rest assured that I am truly grateful and the dark secrets of your ignorance are safe with me.

I must, however, thank Richard Milbank at Atlantic Books, for his unflagging interest in this project, his tolerance of my authorial foibles and the immense help he has given throughout in steering this work from concept to completion. I am also indebted to my editor, Ian Crofton, whose encyclopedia knowledge, and refusal to let me get away with vagueness or feeble attempts at humour have raised the respectability of this book at least a couple of intellectual notches. Any errors or bad jokes that remain, however, are all my own.

BIBLIOGRAPHY

To list all the books and journals I have consulted in the preparation of these 501 unknowns would probably result in a list longer than the rest of the book. So here is a selection of some of the references I have found most useful and interesting, for anyone in search of more detailed information. The references are indexed according to the number of the unknown item.

Introduction: Polak and Rashed (2010). 'Microscale laser surgery reveals adaptive function of male intromittent genitalia', *Proceedings of the Royal Society B Biological Science*

1.   Yang, F. et al. (2003), 'Reciprocal chromosome painting among human, aardvark, and elephant (superorder *Afrotheria*) reveals the likely eutherian ancestral karyotype', *Proceedings of the National Academy of Sciences USA*

9.   Bello, S.M. et al. (2011), 'Earliest directly dated human skull-cups', *PLoS ONE*

24.   Andreasen, R.O. (2000), 'Race: biological reality or social construct?' *Philosophy of Science*

35.   Ellington, C.P., van den Berg, C., Willmott, A.P., Thomas, A.L.R. (1996), 'Leading-edge vortices in insect flight', *Nature*

35.   Buchwald, R. and Dudley, R. (2010), 'Limits to vertical force and power production in bumblebees (Hymenoptera: *Bombus impatiens*)', *The Journal of Experimental Biology*

36.   Muller, H., Grossmann, H., Chitko, L. (2010), '"Personality" in bumblebees: individual consistency in responses to novel colours?' *Animal Behaviour*

37.   von Frisch, K. (1946), 'Die Tänze der Bienen', *Österreichische Zoologische*

37. Su, S. et al. (2008), 'East Learns from West: Asiatic honeybees can understand dance language of European honeybees', *PLoS ONE*

37. Klein, B.A. et al. (2010), 'Sleep deprivation impairs precision of waggle dance signaling in honey bees', *Proceedings of the National Academy of Sciences USA*

44. Rovner, S.A. (2010), 'Recipes for limb renewal', *Chemical & Engineering News*

46. Klem Jr., D. (1989), 'Bird-window collisions', *Wilson Bulletin*

47. Heyers, D. et al. (2007), 'A visual pathway links brain structures active during magnetic compass orientation in migratory birds', *PLoS ONE*

48. Evans, C.S. and Evans, L. (1999), 'Chicken food calls are functionally referential', *Animal Behaviour*

49. Searcy, W. and Beecher, M. (2011), 'Continued scepticism that song overlapping is a signal', *Animal Behaviour*

59. Bickart, K.C. et al. (2010), 'Amygdala volume and social network size in humans', *Nature Neuroscience online*

60. Quian Quiroga, R. et al. (2005), 'Invariant visual representation by single-neurons in the human brain', *Nature*

72. Wells, D.L. and Millsopp, S. (2009), 'Lateralized behaviour in the domestic cat, *Felis silvestris catus*', *Animal Behaviour*

77. McGrew, W.C. and Marchant, L.F. (1999), 'Laterality of hand use pays off in foraging success for wild chimpanzees', *Primates*

90. Rogers, P.J. et al. (2010), 'Association of the anxiogenic and alerting effects of caffeine with ADORA2A and ADORA1 polymorphisms and habitual level of caffeine consumption', *Neuropsychopharmacology*

95. Neumayr, A. (1994), *Music and Medicine*, Vol. 1, Medi-Ed Press, Bloomington, Illinois

101. Damasio, A. (2010), *Self Comes to Mind: Constructing the Conscious Brain*, William Heinemann

111. Ellis, G.F.R. and Uzan, J.-P. (2005), 'c is the speed of light, isn't it?' *American Journal of Physics*

121. Baron-Cohen, S. et al. (1997), 'Is there a link between engineering and autism?' *Autism: An International Journal of Research and Practice*

128. Anandan, C. et al. (2009), 'Is the prevalence of asthma declining? Systematic review of epidemiological studies', *Allergy*

129. Johnson, C., and Eccles, R. (2005), 'Acute cooling of the feet and the onset of common cold symptoms', *Family Practitioner*

135. Rout, T.M., Heinze, D., McCarthy, M.A. (2010), 'Optimal allocation of conservation resources to species that may be extinct', *Conservation Biology*

136. Mascheroni, R.M., Senju, A., Shepherd, A.J. (2008), 'Dogs catch human yawns', *Biology Letters*

187. McKee, A.C. et al. (2009), 'Chronic traumatic encephalopathy in athletes: progressive tauopathy after repetitive head injury', *Journal of Neuropathology and Experimental Neurology*

190. Gelbart, N.R. (2004), 'The blonding of Charlotte Corday', *Eighteenth-Century Studies*

216. Coe, M.J. (1967), '"Necking" behavior in the giraffe', *Journal of Zoology*

222. Nelson, R. (1993), 'Understanding Eskimo science', *Audubon magazine*

222. Lin, R.C. (2004), 'Fractures of the radius and ulna secondary to possible vitamin D deficiency in captive polar bears (*Ursus maritimus*)', *Cornell University senior seminar paper*

223. Levermann, N. et al (2003), 'Feeding behaviour of free-ranging walruses with notes on apparent dextrality of flipper use', *BMC Ecology*

225. Forrester, G.S. et al (2011), 'Target animacy influences gorilla handedness', *Animal Cognition*

228. Gallup, A., Miller, M., Clark, A. (2009), 'Yawning and thermoregulation in budgerigars, *Melopsittacus undulates*', *Animal Behaviour*

228. Gallup, A.C. and Gallup, G.G. (2007), 'Yawning as a brain-cooling mechanism: nasal breathing and forehead cooling diminish the incidence of contagious yawning', *Evolutionary Psychology*

230. Panksepp, J. and Burgdorf, J. (2003), '"Laughing" rats and the evolutionary antecedents of human joy?', *Physiology & Behavior*

232. Miller, G.F. (2000), 'Evolution of human music through sexual selection', in Wallin, N.L., *The Origins of Music*, MIT Press

235. Setchell, J.M. et al (2009), 'Opposites attract: MHC-associated mate choice in an Old World primate', *Journal of Evolutionary Biology*

237. Brown, P. et al. (2004), 'A new small-bodied hominin from the late Pleistocene of Flores, Indonesia', *Nature*.

243. Bailey, N.W. et al. (2010), 'Acoustic experience shapes alternative mating tactics and reproductive investment in male field crickets', *Current Biology*

244. Maye, A. et al. (2007), 'Order in spontaneous behaviour', *PLoS ONE*

256. Mithen, S.J. (2005), *The Singing Neanderthals: the origins of music, language, mind and body*, Harvard University Press

260. Pullum, G. (1991), *The Great Eskimo Vocabulary Hoax and Other Irreverent Essays on the Study of Language*, University of Chicago Press

261. Herndon, J.M. (2007), 'Nuclear georeactor generation of the Earth's geomagnetic field', *Current Science*

264. Brody, S. (1945), *Bioenergetics and Growth*, Reinhold

264. White, C.R. and Seymour, R.S. (2003), 'Mammalian basal metabolic rate is proportional to body mass$^{2/3}$', *Proceedings of the National Academy of Sciences USA*

265. Reznikova, Z. and Ryabko, B. (2011), 'Numerical competence in animals, with an insight from ants', *Behaviour*

269. Crofoot, M.C. et al (2010), 'Does watching a monkey change its behaviour?' *Animal Behaviour*

270. Braithwaite, V.A. (2010), *Do Fish Feel Pain?* Oxford University Press

276. Manger, P.R. (2006), 'An examination of cetacean brain structure with a novel hypothesis correlating thermogenesis to the evolution of a big brain', *Biology Review*

277. Johnson, M.E. and Atema, J. (2005), 'The olfactory pathway for individual recognition in the American lobster *Homarus americanus*', *Journal of Experimental Biology*

277. Aquiloni, L. and Gherardi, F. (2010), 'Visual recognition of conspecifics in the American lobster, *Homarus americanus*', *Animal Behaviour*

282. Kaptchuk, T.J. et al. (2010). 'Placebos without deception: a randomized controlled trial in irritable bowel syndrome', *PLoS ONE*

287. Dawson, W. (1927), 'Mummy as a drug', *Proceedings of the Royal Society of Medicine*

313. Zegers, R.H.C., Weigl, A., Steptoe, A. (2009), 'The death of Wolfgang Amadeus Mozart: an epidemiologic perspective', *Annals of Internal Medicine*

334. Dickey, J. (2003), 'The structural dynamics of the American five-string banjo', *Journal of the Acoustical Society of America*

352. Ali, J. and Huber, M. (2010), 'Mammalian biodiversity on Madagascar controlled by ocean currents', *Nature*

362. Schmidt, H. (1978), 'Can an effect precede its cause? A model of a noncausal world', *Foundations of Physics*

382. Hutchinson, G.E. (1961), 'The paradox of the plankton', *The American Naturalist*

385. Kepr, F. et al. (2006), 'Methane emissions from terrestrial plants under aerobic conditions', *Nature*

385. Nesbit, R.E.R. et al (2009), 'Emission of methane from plants', *Proceedings of the Royal Society B: Biological Sciences*

386. D'Arrigo, R. et al. (2007), 'On the "divergence problem" in northern forests: a review of the tree-ring evidence and possible causes', *Global and Planetary Change*

396. Bauval, R. and Gilbert, A. (1994), *The Orion Mystery*, Cornerstone

415. Cross, F.R. (2009), 'How blood-derived odor influences mate-choice decisions by a mosquito-eating predator', *Proceedings of the National Academy of Sciences USA*

417. Young, L.C. (2008), 'Successful same-sex pairing in Laysan albatross', *Biology Letters*

446. Bercel, N.A. (1959), 'The effect of schizophrenic blood on the behavior of spiders', *Neuropsychopharmacology*

448. Eberhard, W. (2011), 'Are smaller animals behaviourally limited?' *Animal Behaviour*

450. Gaskett, A.C. et al. (2004), 'Changes in male mate choice in a sexually cannibalistic orb-web spider', *Behaviour*

453. Waterman, J. (2010), 'The adaptive function of masturbation in a promiscuous African ground squirrel', *PLoS ONE*

470. Genda, H. and Ikoma, M, (2007), 'Origin of the ocean on the Earth: early evolution of water D/H in a hydrogen-rich atmosphere', *Icarus*

481. Schärer, L. et al. (2010), 'Mating behavior and the evolution of sperm design', *Proceedings of the National Academy of Sciences USA*

499. Grant, S. (2009), 'From little things, big things grow', *Sydney Alumni Magazine*

INDEX